T0315308

The Biology of Us

The Biology of Us

The Living World All Around and In Us

GARY C. HOWARD

OXFORD
UNIVERSITY PRESS

OXFORD
UNIVERSITY PRESS

Oxford University Press is a department of the University of Oxford. It furthers
the University's objective of excellence in research, scholarship, and education
by publishing worldwide. Oxford is a registered trade mark of Oxford University
Press in the UK and certain other countries.

Published in the United States of America by Oxford University Press
198 Madison Avenue, New York, NY 10016, United States of America.

© Oxford University Press 2024

CIP data is on file at the Library of Congress

ISBN 978–0–19–766479–7

DOI: 10.1093/oso/9780197664797.001.0001

Printed by Marquis Book Printing, Canada

For Shirley.

Contents

Tides of the Body
by Anne Whitehouse

Breath, shape-changer,
the organs gently swaying in their fascial hammocks
like the flora and fauna of an undersea world—
the yellow of the small intestine,
deep coral of the liver, green bile duct,
pancreas the color of the ocean floor.
Blood circulating through arterial rivers
in an endless loop.

Gently I placed my fingers
over the openings of my ears.
The sound of my breath inside my throat
was like the echo in a seashell,
ever-present, softly audible.
I tuned out the world for a moment
so I could listen.

Preface

Sadly, nature and biology are easily overlooked. Our jobs, families, activities, social media, and more leave little time for thinking about biology. While our ancestors had to hunt and gather food and protect themselves from wild animals, we hunt for bargains and gather food at the supermarket. Our interactions with wild animals consist mostly of complaining about mosquitoes at a barbeque. We are protected from the elements with air conditioning in the summer and heating in the winter. Only major life events, such as birth, injury, disease, and death, remind us that we are still biological organisms. The life all around us is invisible.

When we do "interact" with nature, it is often in a controlled and safe environment, such as a zoo or theme park, or on television. Several beautiful television shows, films, and books have inspired people to appreciate science. However, the setting is nearly always an exotic location (e.g., the Kalahari Desert, the deep ocean, or Antarctica). These are treasures. They bring the wonder, joy, and tragedy of nature to huge audiences. But they also inadvertently reinforce our separation from nature. Biology happens somewhere else.

Yet, biology is in plain sight in us and all around us. Amazing species live with us in our backyards and in and on us. Many coexisted with or even preceded the dinosaurs. This book will describe, for a lay audience, the common but fascinating examples of biology and nature that are hidden in plain sight in our daily lives. It will focus on human biology, but describe animals and plants all around, on, and in us to put human features into an evolutionary context. Many aspects of ourselves and our normal activities are examples of evolution: breathing, eating, standing up, communicating, telling time, and more. The book will illustrate evolutionary strategies used successfully by common organisms for hundreds of millions of years. If we notice those organisms at all, we see them as trivial neighbors or even pests, but they are just as amazing as those in the Serengeti or the Galápagos Islands.

The recent tragic pandemic brought us all face to face with biology and reminded us that we humans are animals too and part of the natural world. Furthermore, we humans have an affinity for nature. E.O. Wilson called it "biophilia," and psychologists have shown convincingly that exposure to nature is beneficial to human emotional and physical well-being. This book seeks to increase awareness of biological phenomena that are all around us all the time. Hopefully, the reader will come to notice and enjoy nature better.

This book is the result of several years of work. Many people helped me with it, mostly by teaching me about science and how to appreciate the workings of nature. I have been fortunate to work with and around multiple outstanding scientists at many institutions. Specifically, I thank my primary mentors: Ramsey Frist (West Virginia University); John Mayfield, John Ellison, and William Brown (Carnegie Mellon); and Sarah C.R. Elgin (Harvard). Willis Hertig (plant morphology) and Henry Hurlbutt (invertebrate zoology) would be surprised to learn how their classes at West Virginia University have stayed with me over the years. I also thank Jeremy Lewis for his support and patience and all of the extraordinary team at Oxford University Press. Finally, I thank my wife, Shirley, for her unending patience and encouragement throughout.

1

Looking for Sowbugs

My young children and their friends used to search our backyard for "rollie-pollies." They would look under logs, in the leaf litter, beneath flowerpots or outdoor pet dishes, and under paving bricks or stones, and they would enjoy watching these harmless little animals roll up in their hands. The children had no idea that these delicate creatures were crawling around long before the dinosaurs ruled the earth. The same is true for many other plants and animals that we see every day.

Rollie-pollies have lots of names: pillbugs, sowbugs, or woodlice, and they are fascinating animals (Schmidt, 2008). As delicate as they are, they live about two years but can reach five. Sowbugs (suborder Oniscidea) are crustaceans (Broly et al., 2013) and the only crustaceans that live their entire lives on land (Figure 1.1). Although seemingly simple and fragile, the rollie-pollies have enjoyed a very long existence on earth. They crept out of the sea and onto land more than 300 million years ago or 75 million years before the first dinosaurs appeared. Today there are nearly 4,000 known species of rollie-pollies and many more unknown. They have a calcareous exoskeleton and jointed appendages, and so, they are more closely related to crabs and lobsters than to insects. Females have a marsupium or brood pouch, somewhat like a kangaroo. The eggs incubate in the marsupium until they hatch, and the family group remains together until the young are mature.

Unlike most of their close relatives in the order Isopoda that stayed in the water, rollie-pollies crawled onto the land, and in doing so, they had to make significant changes to survive. First, they had to adapt to a nonwater environment. They had to evolve a way to obtain oxygen from the air rather than the water. They accomplished this with pleopods, appendages on the underside of the abdomen that contain respiratory structures that enable gas exchange. Nevertheless, moisture remains important to them, and so they live where it is moist and shaded. Next, they had to adapt to handle water and waste products. A water duct system also serves to eliminate nitrogenous waste. As the liquid waste products move along the duct system, they become more

The Biology of Us. Gary C. Howard, Oxford University Press. © Oxford University Press 2024.
DOI: 10.1093/oso/9780197664797.003.0001

Figure 1.1. Rollie-Pollie (Family: Armadillidae).
Reproduced from Wormsandstuff (2015). Armadillidae, a pill bug, doodle bug, potato bug, roly-poly. https://commons.wikimedia.org/wiki/File:Pill-bug_doodle-bug_potato-bug.jpg. Public domain license (CC0).

alkaline, and the ammonia-containing waste compounds are converted to nitrogen gas that is expelled.

Rollie-pollies are extraordinary creatures, but they are small and common and essentially invisible in that they are simply overlooked. They are only one of many amazing species that we never see. Many more are all around us and in and on us. Furthermore, we are embedded in that natural environment. No matter how many modern conveniences and devices that we surround ourselves with, we cannot escape that natural world that molded us through evolution. That evolution continues today.

Evolution concerns strategies used by species to survive. Much of it involves simple solutions to the problems of day-to-day survival, and these translate to the varied solutions to a series of engineering challenges. The easiest to see are the ways in which living organisms "stand up." Kelp developed nodules that fill with air and float the kelp in the ocean. When plants moved onto land, they evolved other means of supporting themselves so that the process of photosynthesis could be optimized. In like manner, animals have developed external or internal skeletons to support them in an upright manner. Other engineering questions similar to these are solved right down to the molecular level.

Over the last four billion years or so, a dazzling number of species have evolved. Stephen J. Gould described the Burgess Shale, a geologic formation with an extraordinary collection of fossils from the Cambrian Explosion (Gould, 1990). Nearly all of those life forms were lost in a mass extinction, but the shapes and designs of those and of later organisms are almost beyond imagination. In fact, if a creature can be imagined, it has almost certainly already lived on Earth somewhere and some time. Tiny bacteria and huge bison, green algae and African elephants, wood ducks and dodos and dinosaurs. Beautiful or ugly. Large or small. And everything in between. Life on Earth remains amazingly diverse. It is all around us, and we are a part of it.

Moreover, although we may not always realize it, we all have an innate affinity for nature and biology. E.O. Wilson (1997) named it "biophila." He suggested that all humans have a strong affinity for other living organisms. His concept of biophilia is supported by numerous studies in environmental psychology showing the emotional and physical benefits of interactions with nature. Kuo (2015) suggested that adding green spaces to areas with high health risks would be an inexpensive public health intervention. Nature is restorative on many levels (Kaplan, 1995). It improves mood and vitality, reduces stress and anger, and speeds recovery. Many people sense this on some level. In an excellent review, Gifford (2014) notes 10 reasons why individuals engage in nature activities: "They believe that nature activities will facilitate a sense of cognitive freedom, allow them to simply experience nature, enhance their ecosystem connectedness, escape from stress, offer a physical challenge, foster personal growth, provide an opportunity to guide others, heighten their sense of self-control, renew social connections, and improve their health" (p. 559). Interactions with nature yield a positive, proportional decrease in stress and reductions in blood pressure, heart rate, and levels of cortisol and markers of inflammation (Mitchell and Popham, 2008; Tsunetsugu et al., 2010). A prospective randomized trial found that nature activities were effective in reducing stress in the low-income parents of school-age children (Razani et al., 2018).

Sadly, it is too easy to overlook nature. We simply take it all for granted. It's understandable. Our lives are a blur. Children, friends, work, school, television, social media, and other activities and responsibilities consume our lives. Who could have time to notice anything? Many reasons it is easy to overlook nature. Here are a couple.

First, we modern humans (at least those of us fortunate to live in the developed nations) are almost completely insulated from nature. We live in comfortable houses that protect us from the rain and snow. Storms are a more of an inconvenience than a threat. We adjust the temperature with a few taps on a small computer screen that controls the heat or air conditioning. Our ancestors might have been hunter-gatherers, but now, our hunter-gathering activities are confined to the local supermarket. We venture out into the "wild" to mow the lawn or water the flowers. The only wild animal we have to face is an occasional mosquito at a barbeque. Without a day-to-day struggle with nature, we don't have to know much about it. We can ignore it. However, despite cappuccinos and dot.coms, iphones and immunizations, we are as much a part of nature as we ever were. The dirt under our feet is the same dirt that was walked on by dinosaurs and wooly mammoths. And it was old even when they were here. Hunger and pain, cold and heat, bee stings, cuts, broken bones, childbirth, pandemics, and death rapidly bring us face to face with our biological selves and remind us that we are still animals.

Second, even when we do think about nature, we typically think of biology as something that happens somewhere else. We might see a bit of nature on a vacation to the beach or while camping at a national park. That sense that nature is somewhere else has also been reinforced by a number of wonderful television programs that have brought the extraordinary diversity and beauty of life into our homes. These beautiful descriptions tend to emphasize exotic life forms around the world. They show the struggles of penguins in the frozen Antarctic or a family of lions in the Serengeti. David Attenborough has been particularly successful. His programs are beautifully photographed and intelligently narrated. Other nature programs have followed his lead, and now, whole channels featuring the wonders of biology have sprung up, including *Discovery* and the *National Geographic*. These movies and programs build on a rich tradition. Even Darwin, whose name is so associated with evolution, traveled in the ship *Beagle* and famously spent time in the Galapagos Islands to study animals and plants. Alfred Russel Wallace, the codiscoverer of natural selection, studied in the Amazon River basin and later in the Malay Archipelago. They worked independently, but published their findings at the same time. Thus, it is too easy to think of biology and evolution as happening someplace other than where we live.

Yet, whether we notice or not, the fact remains that the biological world is still all around us. We don't need to go to Mount Kilimanjaro or the Amazon or the Arctic tundra to see amazing plants and animals that demonstrate the inventiveness of evolution. Equally amazing animals and plants are everywhere with us, no matter where we are. The rollie-pollies are here to remind us that we are still part of nature and that the natural world is all around us. They and other animals, mostly unseen, thrive alongside us in cities and suburbs.

Even when we see nature, we might misinterpret it. A robin in the early spring scratching in the leaf litter reminds that spring has arrived and that

Figure 1.2. Robin (*Turdus migratorius*). A robin scratching in the leaf litter in the back yard might be seen by us as a welcome harbinger of spring. However, to an earthworm, it might look more like a velociraptor.
Photograph reproduced with permission from Rebecca Howard Valdivia.

realization gives us a warm feeling of peace and serenity (Figure 1.2). But to the worm desperately trying to hide, the robin looks a lot like its ancient feathered relative, the velociraptor (*Velociraptor mongoliensis*) (Turner et al., 2007). To the worm, the world is the same as it was 150 million years ago, and life is a constant struggle to avoid the robin and other predators. We humans love nature, but we are rarely threatened by it and certainly not on a daily basis in the way that nearly all other organisms are.

The diversity of life is astounding, and equally exotic species live right under our feet. In fact, evolution and major aspects of biology can be seen anywhere. All around us, we can see trees, flowers, birds, people, mushrooms, insects, and many others. In fact, it is hard to find a place that is not touched by some form of life. Living organisms fill essentially every niche and often compete for the same niche. The beauty of Darwin's "natural selection" is that the fittest organisms—those that can reproduce most effectively—survive (Marshall, nd). Others are lost. That mechanism explains how organisms have evolved over time, and it underlies all of our understanding of biology and our knowledge of how organisms work and how disease affects them and us. The result of evolution is a world filled with uncounted numbers of different organisms.

Yet even in this vast diversity, nonbiologists can see similarities and differences in organisms. We can easily see that a rounded lobes of a white oak leaf differ from that cut forms of a black oak leaf. However, both produce acorns so we can recognize them as related on some level. We can also see that lions, tigers, panthers, and house cats share the general features of cats. We would not likely put a robin or trout in that group.

Humans have carefully looked at plants and animals for millennia. For most of that time, simple survival depended on it. Our ancestors had to know which berries were safe to eat and which were not. They had to recognize plants with roots and tubers that contained water. They passed that knowledge from generation to generation by oral tradition. At some point, particularly in India and China but also later in Greece, humans began to be interested in the organisms themselves; they started to group them according to their characteristics and invented names for most of them.

Modern biologists have taken the classification to the next level and applied very careful analysis to try to determine what animals and plants go together and how they are related or not. The first step in understanding living organisms was to know what organisms there are, a sort of biological inventory. Early biologists tried to organize living creatures by their characteristics, but it was Linneaus who established the two-name system that we still used to day. Humans are *Homo sapiens,* which translates literally to "wise man." The term *sapiens* is our species, and *Homo* is the genus, which is a group of related species. Species is the lowest level of differentiation. Its definition is a functional one: two organisms of a single species can produce fertile off-spring. If they cannot do that, then they are two different species.

Scientists added additional levels of complexity to the system. These include family, order, class, phylum (for animals) or division (for plants), and kingdom (Jezkova and Wiens, 2017). The differences between species can be quite subtle, but those between the higher levels of organization are much more dramatic.

The driving force behind these differences, as noted above, has been the survival of the fittest. Surviving each challenge required the solution of an engineering problem, and the structures and behaviors that we see in different organisms now are those solutions. In fact, evolution can be seen as a sum of these engineering solutions. So living organisms, including humans, are collections of engineering solutions to specific problems in biology.

This engineering concept also applies to us humans. Everything we are or do depends on biological systems that we take for granted. We get up in the morning. Breathe. Eat. Move. Our ability to do those things is based on bones, muscles, nerves, organs, blood vessels, lungs, and many other cellular, molecular, biochemical, and physiological systems, all of which evolved over millions of years. Each new "invention" added new capabilities. Over time, the sum of these changes and their effects on the ability of organisms to reproduce resulted in a new species.

This book will describe common but fascinating examples of biology and nature that are hidden in plain sight in our daily lives. It will focus, in particular, on one amazing animal that we see every day: humans. Humans are not the strongest or fastest, but we have managed to spread across the entire world. Many aspects of ourselves and our normal activities are examples of evolution: breathing, eating, standing up, communicating, telling time, and more. The book will illustrate evolutionary strategies used successfully by common organisms for hundreds of millions of years. If we notice those organisms at all, we see them as trivial neighbors or even pests, but they are just as amazing as those in the Serengeti or the Galápagos Islands. The book will begin outside of the human body and work its way more or less inward.

Life is astoundingly diverse. Hollywood would have a difficult time imagining organisms as amazing as those that nature has already experienced. Many of these are marine or extinct, but it is quite easy to see several of the major strategies that life uses in the creatures we walk past every day. Life evolved by solving a series of engineering problems to take advantage of new ecological niches. Chapter 2 will review the characteristics of the major groups and provide common examples of them. Examples include the very primitive land planaria (Platyhelminthes), earthworms (Annelida), slugs and snails (Mollusca), ants and daddy longlegs spiders (Arthropoda), and birds, fish, and us (Chordata).

After a rain, the air takes on a particular smell. Bacteria in the soil, actinomycetes, secrete organic compounds, including geosmin and 2-methylisoborneol. The soil contains many more important organisms and processes (Chapter 3). It's easy to overlook what is under our feet, but our lives would not be the same or even possible without the processes that take place in the living world lies right beneath our feet. Nitrogen is fixed for a number of key chemicals (e.g., amino acids, nucleic acids, chlorophyll, and energy molecules [ATP, NAD]). Seeds germinate. Decomposition occurs, and 80%–90% of plants have symbiotic relationships with fungal rhizomes that are critical to the uptake of nutrients into plants.

Many living organisms are hidden in plain view all the time. One of the most interesting is the supermarket. They contain large collections of organisms, living or previously living. In fact, a supermarket is a great place to learn about biology. There are more examples of plant divisions and animal phyla organisms (e.g., marine Mollusca and Arthropoda). The vegetables, fruits, and fungi demonstrate features of those organisms. The "Meat" section of the chapter includes organs and tissues (e.g., muscle, bone, cartilage, heart, kidney, liver, stomach [tripe], brain) that are similar to those of humans. Even better, at the supermarket, we can take things home and examine them close up and even take them apart. In Chapter 4, we will be looking at our food as a collection of biological specimens.

Lots of organisms are even closer to us. In fact, they live on and even inside us (Chapter 5). They include the pests that visit us, such as fleas and ticks, and the bacteria that encrust our teeth. They also live inside us. Some are parasites and some are needed for our health. In addition, we also carry the "remains" of viruses that infected humans thousand and millions of years ago.

The idea of inside and outside seems simple, but it's one of the most basic engineering problems that living organisms had to solve (Chapter 6). And once an inside and outside have been established, we need a mechanism to pass materials and information into and out of the organism.

Humans also have an inside and an outside, although it might not be exactly what one imagines. Topology is the mathematical study of shapes, and we humans are topologically a torus. We have an opening at our mouths and another at the anus, and in between, is a long tube. The inside of the alimentary canal is actually the outside of us. In Chapter 7, we will discuss the inside and outside of humans.

Eating is a fundamental biological process for every animal, including us. Sometimes we eat, and sometimes we are eaten. In Chapter 4, we looked at the foods in our supermarkets as examples of living organisms in our environment. In Chapter 8, we look at what we eat. For omnivores, humans are fairly picky eaters in that we focus mostly on vertebrates and a few plant divisions. We will also explore the anatomy, physiology, and process of eating. Food is also intertwined with our evolution. Our teeth and those of our ancestors reflect those changes as we moved from hunter-gatherers to early agriculture. The introduction of cooking allowed us to get more nutrition from our food. Our relationship with food has recently been changing. For millennia, our ancestors struggled to get enough food, and we developed genetic processes to extract as much value from our food as possible. Now for most in the developed world, we are awash in calories and are suffering diseases of affluence (e.g., obesity, heart disease, diabetes).

We all stand up many times a day (Chapter 9). Although we hardly think about it, that movement requires the coordination of many muscles, a solid skeletal system, a keen sense of balance, and appropriate control of blood pressure. Interestingly, other organisms have developed different strategies for holding themselves up. Trees and plants use roots implanted in the soil and build their structures from cellulose and lignin to grow up toward the light and cause their roots to grow down in response to gravity. Standing up and moving are critical for human health. A sedentary existence (being a "couch potato") has a high cost in terms of obesity, diabetes, heart disease, and more.

Blood is a miraculous fluid (Thiebes, 2021). It carries oxygen and nutrients throughout our body. It also carries away waste products and carbon dioxide. As if that were not enough, blood transports signaling molecules (e.g., hormones), salts, buffers, and lipids. It also transports the various components of the immune system. Finally, if a leak occurs, blood has its own self-sealing chemistry in the coagulation system. We will look carefully at blood in Chapter 10.

Death is usually seen as the end of life, but one of the most astounding facts about death is that every human cell and every other cell that we know of includes genetic death programs that are activated for specific purposes (Chapter 11). For humans, those include the development of fingers and toes. Our adaptive immune system works as well as it does because programmed cell death eliminates cells that make antibodies that bind to our own tissues. These programs underlie diverse processes in plants and animals. In metamorphosis, the tails of tadpoles disappear from cell death as the frog's legs develop, and caterpillars change dramatically to become butterflies. Leaves and ripe fruit fall because a layer of cells dies at the right time to release them. Even bacteria commit suicide when the colony has outstripped its resources. The cells communicate with each other by secreting chemicals in a process called quorum sensing.

The human ability to communicate through speaking and writing is thought by some to be one of the characteristics that distinguishes us from other animals. However, we also communicate with neurotransmitters, hormones, cytokines, and chemokines and other factors (Chapter 12). Our senses involved the communications of information across a membrane, and secondary messengers within the cell and nucleus that turn genes on and off. Most of our drugs use the similar types of proteins called G-protein-coupled receptors. During development, tissues communicate to ensure that organs take on their correct size and shape. Neurons use neurotransmitters and electrical signals to communicate with each other and our muscles.

One of the great remaining problems in biology is to understand the relationship between the physical brain and consciousness (Chapter 13), and also the relationship among sleep, consciousness, anesthesia, and coma. For example, we all need sleep, but its purpose is still not understood. We know that the lack of sleep has serious consequences: our immune system is less effective, our mental acuity is degraded, and more. And when we are sleeping, we are relatively defenseless. Why would evolution leave us with this need? The benefit must be enormous to overcome the seemingly negative pressures against sleep.

Most of us take it for granted that we wake up in the morning after a good night's sleep and get sleepy again in the late evening. Many physiological processes in humans, other animals, and even some bacteria obey a regular pattern based on a 24-hour solar day. These circadian rhythms are the most obvious way we "tell time," but there are others, such as hormonal cycles and sleep phases (Chapter 14). For example, human development involves complex programs of genetic pathways that must be turned on or off at the proper times, and aging involves many unanswered questions.

We have come a long way in understanding biology (Chapter 15). To a great extent, we now control our fate. We have extended the human lifespan. We regulate our immediate environment with heating and air conditioning. Our food supply is relatively secure. We have mitigated the effects of evolution with eyeglasses, hearing aids, and other devices. The climate crisis will likely change that. The excessive heat will put pressure on water and food supplies. Some regions might become uninhabitable. Populations will be forced to move. Those radical environmental changes will result in natural selection.

In summary, this book seeks to remind us that we are part of biology and nature. Even though our modern comforts insulate us well against the natural phenomena that our ancestors struggled with, we are still part of evolution. Reminders are all around us if we just take a few moments to savor them. They are as simple as dew drops, spider webs, and the soil under our feet. They also manifest themselves in out ordinary day-to-day activities of eating, walking, and sleeping. So let's get going and take a look at some of them.

2

Life around Us

Amazing life is all around us. We just have to look. An adult monarch butterfly (*Danaus plexippus*) weighs only half a gram on average, less than a penny. Yet, these tiny, fragile insects migrate thousands of miles each year (Reppert and de Roode, 2018). Those in the eastern and central United States overwinter in Mexico. Butterflies west of the Rockies migrate south to California. In either case, it's an incredible journey. A terrestrial flatworm hunts among the leaf litter for earthworms and other prey. But if it is cut in half, each half will regenerate a whole worm. Often in the summer, roadside ponds turn green with pond scum that contains a virtual Jurassic Park of microscopic plants and animals. These and myriad more plants and animals live all around us. We walk past them every day but don't see them. They are hidden in plain sight.

Trees, flowers, birds, worms, mammals, mushrooms, insects, and many other living things cover most of the planet. In fact, it is hard to find a place that is not touched by some form of life. Living organisms fill essentially every niche and often compete for the same niche. They use different strategies to survive. Whether they are lions in the Serengeti or earthworms in our garden, they are all extraordinary. Here we will focus on the life all around us right in our own homes, backyards, and neighborhoods.

Where to start? The vast number of different living organisms can be a little overwhelming. It's hard to see the trees for the forest, so to speak. How can we make sense of the many different plants and animals that are all around us? What should we look at? White and black oaks both produce acorns, so they must be related, but their leaves are quite different. Cats and dogs have similarities, but have enough differences so they are easy to tell apart. A butterfly would likely not be confused with a worm. But what about a caterpillar? And how would the caterpillar and worm fit with a centipede?

Fortunately, biologists have worked out many of those relationships. Knowing how they see the different groups will help us to better appreciate the many living things that we live with. Moreover, we will better appreciate those differences if we review the basics of taxonomy. Biologists call

The Biology of Us. Gary C. Howard, Oxford University Press. © Oxford University Press 2024.
DOI: 10.1093/oso/9780197664797.003.0002

this systematics. Each living group solved major engineering challenges (i.e., standing up, reproducing, moving, and exchanging gases) to survive in a tough world. After all, every living organism on Earth uses the same DNA/RNA to encode genes, and that simple fact implies that we all came from the same lineage. We are all related, but over hundreds of millions of years, evolution created an astounding diversity of plants, animals, and more. Furthermore, we, the ants, flowers, and all other living organisms have survived countless assaults for millions of years and even several mass extinctions in which nearly all life on Earth died. We are the survivors (for now).

Building the Family Tree

Humans have been classifying plants and animals since we began communicating. Early humans had to know which plants were good to eat and which were toxic. So they invented common names for most of them and grouped organisms according to the features they could see. Early biologists also tried to organize living creatures by their characteristics. However, the real breakthrough came from the Swedish scientist Carl Linnaeus, who established the two-name system that we still use today. His masterpiece *Systema Naturae* was first published in 1735, but the 10th edition of 1758 is generally cited as the real beginning of modern systematics. He was the first to use the binomial system of Latin names for the genus and species of each organism. Thus, humans are *Homo sapiens*, which translates literally to "wise man." The term *sapiens* is our species, and *Homo* is the genus, which is a group of related species. We are all indebted to him for his contributions, and no one recognized those contributions more than Linnaeus himself. He noted, "No one has been a greater Botanicus or Zoologist" (Bibell, 2008). In spite of his "modesty," it was a singular achievement in science.

Since Linnaeus, scientists have continued to refine the family tree. Biologists apply very careful analysis to try to determine which animals and plants go together and how they are related or not. In this way, they hope to bring order to the seeming chaos of life. They added additional levels of complexity to the system, including family, order, class, phylum (for animals) or division (for plants), and kingdom. Species is the lowest level of differentiation. Its definition is a functional one: organisms of a single species can breed to produce fertile offspring. If they cannot do that, then they are two different species. The differences between species can be quite subtle, but those between the higher levels of organization are much more dramatic.

The driving force behind these differences has been natural selection (i.e., the survival of the fittest). Surviving each challenge of life required the solution of an engineering problem, and the structures and behaviors that we see in different organisms now are those solutions. From the beginning, new mutations allowed some individuals to survive and reproduce more effectively. With those mutations, one or another of the many problems of existence was solved in different ways. For example, for millions of years, life remained in the oceans. Life in the water had its own challenges, but the transition to living on land brought new engineering problems involving gas exchange, reproduction, movement, and more. As evolution proceeded, further mutations afforded some groups new ways of dealing with those challenges. In fact, evolution can be seen as a sum of these engineering solutions. So living organisms are collections of engineering solutions to specific problems in biology. This engineering concept also applies to us humans. Everything we are or do depends on biological systems that we take for granted. We get up in the morning. Breathe. Eat. Move. Our ability to do those things is based on bones, muscles, nerves, organs, blood vessels, lungs, and many other cellular, molecular, biochemical, and physiological systems, all of which evolved over millions of years. Each new "invention" (or more correctly, lucky mutation) added new capabilities. Over time, the sum of these changes and their effects on the ability of organisms to reproduce resulted in a new species.

The new structures and/or functions added to an organism as it evolves are the characteristics that biologists use to add them to our family tree. Taxonomy is difficult. It requires careful study of the morphology of various parts of an organism and stages of development. Examination of morphology can sometimes be subjective. In recent years, molecular biology has provided new insights into relationships. New molecular strategies might shed new light. For example, Hebert et al. (2003) suggested the use of a "barcode" for classifying animals. They used the mitochondrial gene for the enzyme cytochrome c oxidase subunit 1 (COI). COI has plenty of sequence changes, which are restricted to the mitochondrial DNA. Hebert et al. found that sequence differences in COI could be used to differentiate closely related species in all phyla except Cnidaria. Laumer et al. (2019) took a different molecular strategy. Instead of focusing on a single gene, they looked at genomic DNA sequences. Their analysis confirmed many of the traditional relationships but also revealed some surprising results and also areas that need additional work. Most importantly, they showed the power of these techniques in systematics.

However, in the final analysis, molecular biology has not proved a universal panacea. There are many reasons. DNA technologies continue to improve, but challenges remain. The fossil record is incomplete. Soft tissues are long lost in extinct species. Still, the assignments remain under intense study, and changes to the classifications occur regularly as more data are obtained and vigorously debated.

However, when thinking about the family tree, we must guard against two prejudices. First, an invention that allows a new species to do something better does not mean that the way the old species does it is not viable. As we will see, many of the "early" models are still flourishing, and they have themselves innovated on that old design to make it even more effective. For example, the insects were here long before the dinosaurs, and they are still doing fine. In fact, insects are easily the most diverse animal group on Earth. There are 6–10 million different insect species, more than any other class of animals.

Second, we humans are not necessarily more complex than earlier more primitive organisms. Maybe. But another interpretation is that single-cell organisms have to do everything (respiration, convert energy, reproduce), whereas multicellular organisms have a division of labor that allows groups of cells to specialize and do less (in a sense). Thus, the momentum toward complexity is in the organization rather than in the number of reactions per cell.

Animals

The animal kingdom is immense and diverse. Animals are all around us, and many are familiar to us. Dogs, cats, hamsters. Pet fish. Pet birds and wild birds. These all have a backbone and fur, feathers, or fin, but they make up only a very small percentage of the animals we live with. Most animals lack a backbone, and some of those are also recognizable. For example, honey bees get good press. Butterflies are beautiful, and many people are planting milkweed to save the monarchs. However, most animals are not nearly as pretty or cuddly as a puppy or kitten. There are no campaigns to save the flatworms.

Table 2.1. Characteristics of the Animal Phyla

Phylum	Examples	Symmetry	Coelom	Segmentation	Int/Ext Skeleton	Backbone
Porifera	Sponges	none	no	no	none	no
Cnidaria	Jellyfish	radial	no	no	none	no
Echinodermata	Sea stars	radial	yes	no	none	no
Platyhelminthes	Flatworms	bilateral	no	no	none	no
Nematoda	Nematodes	bilateral	no	no	none	no
Annelida	Earthworms	bilateral	yes	yes	none	no
Mollusca	Clams	bilateral	yes	yes	external	no
Arthropoda	Insects	bilateral	yes	yes	external	no
Chordata	Humans	bilateral	yes	yes	internal	yes

And cockroaches and spiders are right down at the bottom of the list. Yet even the most disgusting has a great story.

All of those animal groups are classified by the strategies or features that they have evolved to live successfully (Table 2.1). The term "strategies" is not meant to imply intent. Evolution proceeds by natural selection. Mutations result in new strategies that allow an organism to better cope with its specific environment and, thus, reproduce more successfully. That does not necessarily mean that the previous strategies were bad in every environment. Some organisms still do well with the most primitive mechanisms. The features that define the phyla of today include body shape or symmetry, segmentation, coeloms, support structures, and body temperature. Some features are obvious and easy to see. Others are less so.

The most basic feature is body shape or symmetry. Except for the sponges, all animals have symmetry. A simple way to think of symmetry is to ask, Is there an axis of symmetry? In other words, can you cut the animal in two so that the two halves are mirror images? Sponges have no symmetrical organization. Sea stars and jellyfish have radial symmetry: any cut will yield two mirror images. However, higher animals, such as humans, have bilateral symmetry. We have "left" and "right" sides. There is only one axis of symmetry through us.

A second basic feature is segmentation (Figure 2.1). Segmentation is essentially the "copy and paste" application for genetics. Segments are repeated units that make up an animal. The segments might be exactly the same or they might have different features. For example, segmentation can be easily seen

Figure 2.1. Millipede shows segmentation. A millipede at Petroglyph National Monument.

Reproduced from Olsen, A. (2005). Desert millipede (*Orthoporus ornatus*). U.S. Geological Survey. https://www.usgs.gov/media/images/desert-millipede-orthoporus-ornatussw. Public domain license.

in some organisms. Most of the segments of millipedes (Arthropoda) have two pairs of legs, but segments at the head and tail have other appendages. Humans also have segments during development, but our segments are different and more subtle.

Animals also differ in the number of cavities, or coeloms, in the body, and these are related to the different cell layer origins. Animals with bilateral symmetry and a coelom have three tissue layers (i.e., ectoderm, endoderm, and mesoderm). Most animals with radial symmetry (starfish, jellyfish) lack a mesoderm. Sponges have only one type of tissue. A coelom is a space between the intestinal tract and the body wall—that is, a cavity between the outside layer of cells and the layer of cells that form the digestive track of an organism. For example, earthworms are shaped as an elongated cylinder within a cylinder. The cavity between the two cylinders is the coelom. It isn't easy to see the coelom in earthworms or other creatures without cutting them open, but humans are coelomates, and we can much more easily imagine it in ourselves. The coelom begins during early development and divides into several cavities. Again we can't see it under normal circumstances, but its parts can be roughly understood to include the cranial and spinal cavity, the pleural cavity, and the abdominal cavity. Most of our organs are positioned within the coelom.

The stages in some animals (e.g., humans) cannot easily be seen because they occur during embryonic development (Marshall, 2009). The fertilized egg undergoes cell division to increase the number of cells. The inner cell mass forms a ball of cells that invaginates to form a new ball with three layers of cells and a hole where the invagination began. Imagine a deflated basketball that is pushed in on one side to form a sort of half sphere with two layers of rubber. The fate of that hole, called the blastopore, is the differentiating feature in the two large groups. In deuterostomes (e.g., vertebrates), it becomes the anus. In protostomes (e.g., worms, insects, and clams), it becomes the mouth.

Movement also requires engineering strategies. For example, earthworms have no hard support structures, but they have muscles or contractile tissues that allow them to move efficiently. Other animal groups evolved skeletons that are internal or external and serve a variety of functions. Snails have a heavy shell made of calcium compounds that offers protection against predators and helps to prevent them from drying out. In some ways, snails look like an earthworm with a heavy shell. They both leave a slime trail as they move along, and they are both segmented. But snails are mollusks, a phylum that includes clams, oysters, octopus, squid, and slugs. Snails and slugs are members of the class Gastropoda (stomach-foot). Octopus and squid are in the class Cephalopoda (head-foot). The name "snail" refers to both land and sea snails, a mollusk with a shell. Most slugs lack the shell, but look a lot like snails. In fact, some slugs have a small vestigial shell. There are lots of different kinds of snails, and they live on land and in both salt and fresh water. They have either lungs or gills, but amazingly, some land snails have gills, and some fresh water snails have lungs. The shell is made of chitin and calcium carbonate, and it can be expanded as the snail grows by adding more material to the shell.

Mollusks, including snails, have three hallmarks: a mantle, a radula (a file-like structure with thousands of microscopic "teeth"), and a nervous system. The mantle is quite muscular and comprises a significant cavity that contains the gills (or lungs), anus, nephridiospores (kidney-like organs), and the gonopore (the opening for the reproductive organs).

Arthropoda are the largest group of animals on earth. In fact, there are more arthropods than all others combined. The group contains many easily recognizable members, such as crabs, lobsters, shrimp, spiders, scorpions, centipedes, millipedes, sowbugs, and insects. These animals are segmented, but in many, the segments are grouped into other structures. For example, insects have a head, thorax, and abdomen, and lobsters have a cephalothorax and an abdomen). Centipedes are a common arthropod. They have 15–173 true segments. Each has one pair of legs, except the last two and the segment immediately behind the head. The head carries a pair of antennae, and the segment behind the head has a pair of modified legs called maxillipeds or poison claws. Centipedes are predators. They are very fast and eat insects, worms, and mollusks.

In the arthropods, such as sowbugs and insects, the exoskeleton has evolved into a structure of chitin that is more like a suit of armor for a medieval knight. That flexibility was needed for flexible joints. At the joints, the exoskeleton is quite thin and flexible, and that allows for movable joints. Like us, arthropods have "knees" and "elbows" and other joints that bend. The exoskeleton adds protection and a place for muscle attachments, and the flexible joints allow for much greater mobility.

Body temperature is another characteristic that has evolved. Normal human body temperature is about 98.6°F. Even when the ambient temperature is much lower, we maintain that temperature. Like other warm-blooded animals, we use our metabolism to keep ourselves warm. But many animals are cold-blooded. Their temperature is the same as that of the air around them. Today only mammals and birds are warm-blooded. The dinosaurs are thought to have been warm-blooded. In fact, "warm-blooded" and "cold-blooded" are informal terms. Additional variations on body temperature are known, and even among so-called warm-blooded animals, different levels are normal. For example, bats and some birds have greater temperature variations, and sharks maintain a higher temperature in their eyes and brain so they can hunt more efficiently. Body temperature is regulated by the brain, which uses several processes (Morrison and Nakamura, 2019). Vessels in the skin become constricted. Shivering causes muscles to work and produce heat. Metabolism in the brown adipose tissue is inefficient and much of the energy is lost as heat, which warms the body.

Body temperature has some surprising effects. For example, Robert and Casadevall (2009) suggested that the higher body temperatures of many vertebrates protect them against fungal diseases that beset invertebrates.

They found that most fungi do not grow well at mammalian temperatures and estimated that each 1°C increase in body temperature was too much for an additional 6% of fungal diseases. Thus, mammalian body temperature and fevers may be a nonspecific defense against those diseases.

The highest level of classification for animals is the phylum; currently, there are about 31 phyla, although there is disagreement over the total number (Verma and Prakash, 2020). It is important to remember that these systems are human inventions. They are our feeble attempt to understand nature, but nature has its own rules. As scientists learn more about nature, the classification systems become more sophisticated and change. Some of the 31 phyla are quite small. Some are not included here because they are not a typical part of our human environment. Several phyla are completely marine, but some of their members are familiar to us, such as sea stars (Echinodermata) and jellyfish (Cnidaria). Others are mostly parasitic, such as the flatworms (Platyhelminthes) and round worms (Nematoda). Still others are free-living, such as flatworms (Platyhelminthes); roundworms (Nematoda); earthworms (Annelida); slugs and snails (Mollusca); sowbugs, centipedes, and insects (Arthropoda); and vertebrates (Chordata).

Worms

Flatworms (Platyhelminthes) are the simplest animals that we might see. Importantly, use of the term "simplest" is not meant to imply that they are simple or that they are less successful than other groups. Flatworms first appeared on Earth about 550 million years ago. By comparison, the dinosaurs arrived about 230 million years ago and became extinct about 60 million years ago. So flatworms have existed three times longer than the dinosaurs.

If earthworms have dreams, their worst nightmare must be flatworms. Imagine being grabbed and held while your attacker extends his mouth against you, tears into you, and secretes enzymes that dissolve you alive. And even if you somehow manage to rip the attacker in two, each half will regenerate into another flatworm, like some real-life version of the Hydra of Greek mythology.

Many marine flatworms are brightly colored, but others are not cuddly or cute. For those of us in the United States, we are not likely to see flatworms. Nevertheless, they are amazing animals in their simplicity (for an excellent review, see Collins, 2017). They have bilateral symmetry, but lack many of the features of higher animals. They have no real mouth and no gut or coelom. Food enters and exits from the same opening. As their name indicates, they are dorsoventrally flattened. However, they have one thing that lower animals (e.g., sponges) do not: mesoderm. The mesoderm is a tissue layer that, among other things, enables the development of muscle systems and facilitates movement.

While many flatworms are parasitic, there are free-living species (Sluys, 2021). For example, terrestrial flatworms are referred to as Planaria. They need moisture and live in moist areas in the soil, in the leaf litter, under logs, and on plants. All are predators and feed on earthworms, snails, slugs, isopods, larvae, and springtails. Earthworms and other invertebrates are important in the soil, so the Planaria are not welcome invaders. Those engaged in vermiculture are particularly vulnerable to infestations of Planaria. The mouth is positioned in the middle of the body and on the underside. They wrap themselves around their prey and secrete enzymes from their "mouth" that dissolve the prey. Waste left over from digestion must be expelled through the same orifice because they have no anus. Some extend a tube-like mouth out to penetrate the prey. Planaria vary in size from a few millimeters to several centimeters. They tend to be dark in color and can be mistaken for slugs. They have tapered heads, but some also have a "hammerhead." They have a central nervous system and a collection of nerve cells, or ganglion, serves as a primitive brain. Some have eyespots that can detect light intensities. They move by beating cilia that help them glide along on a layer of mucus. They are hermaphrodites. Each animal has both male and female reproductive organs. They also reproduce asexually when animals are divided because they can regenerate so efficiently. They are found in the southern United States, but the shipment of potted plants from one state to another has accelerated their spread. The most amazing feature of planaria is their ability to regenerate lost parts. In fact, they can be cut in half lengthwise or crosswise and regenerate living worms. For this reason, they are a favorite species for research into tissue regeneration.

The most common flatworms for most of us are tapeworms (class: Cestoda). There are 6,000 species of Cestoda, but we most often see *Dipylidium caninum* and *Taenia taeniaeformis*. These appear as what look like grains of rice in the feces or near the anus of cats or dogs. Tapeworms obtain nutrients and oxygen directly through their outer cuticle or tegument. They attach themselves to the intestines of animals by tiny hooks or suckers (depending on the species) on the head or scolex. Behind the scolex is a small neck and then the body. The body consists of numerous segments called proglottids. The proglottids are produced by mitosis in the neck and get larger as they grow. They are packed with eggs, which are released once the segment breaks off and is deposited in the animal's waste. A nerve ganglion in the scolex connects to nerves that run the length of the animal and control muscles that allow the worm to move and to react to touch and chemical signals. The entire worm can attain amazing lengths. For example, *Taenia saginata*, which infects cows and humans, can reach 20 meters (60 feet) in cows. Fortunately, human infections have become rare in recent years. Even cats and dogs have fewer infections. Flea prevention treatments have eliminated the alternate host for the tapeworms.

Nematodes are everywhere (Frazer, 2013). In 1914, Nathan Cobb, a botanist from the Department of Agriculture, provided a colorful description of the number of nematodes in the world.

> In short, if all the matter in the universe except the nematodes were swept away, our world would still be dimly recognizable, and if, as disembodied spirits, we could then investigate it, we should find its mountains, hills, vales, rivers, lakes, and oceans represented by a film of nematodes. The location of towns would be decipherable, since for every massing of human beings there would be a corresponding massing of certain nematodes. Trees would still stand in ghostly rows representing our streets and highways. The location of the various plants and animals would still be decipherable, and, had we sufficient knowledge, in many cases even their species could be determined by an examination of their erstwhile nematode parasites.

In fact, there are 50 to several hundred nematodes per dry gram of soil, but they are not easily seen (Figure 2.2). Most free-living species are essentially microscopic (van den Hoogen et al., 2019). They eat bacteria, algae, fungi, and fecal and dead material. Thus, they are important for soil quality. Those that can be seen are mostly parasitic. *Dirofilaria immitis* causes heartworm in dogs and cats. The worms appear in the feces and vomit. Newborn cats and dogs can be infected through the mother's milk, and all pets need to be "wormed" to protect against these parasites. Several species cause human diseases (Morand et al., 2006). Hookworm infection (*Anclostoma* sp., *Necator* sp.) has become relatively rare in the United States, but pinworm

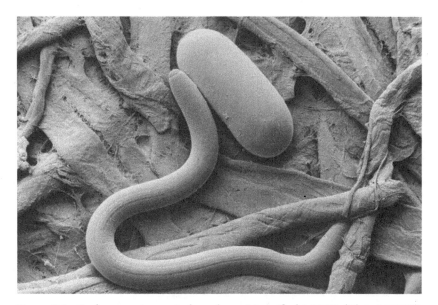

Figure 2.2. Soybean cyst nematode and egg. Magnified 1,000X. (Plate # 10334).
Reproduced from Suszkiw, J. (2009). Bacteria tapped to battle crop-damaging roundworms. USDA Agricultural Research. https://www.ars.usda.gov/news-events/news/research-news/2009/bacteria-tapped-to-battle-crop-damaging-roundworms/.

infections (*Enterobius vermicularis*) are still the most common such infection in children. *Wuchereria bancrofti*, spread by mosquitoes, infects over 100 million people in tropical areas. Once in the lymph system, the worm can grow to 10 cm and causes swellings in parts of the body called "elephantiasis." *Onchocerca volvulus* causes river blindness, a disease most commonly found in Sub-Saharan Africa. Nematodes also infect animals and can be a serious pest in plant crops.

Earthworms (Annelida) are valued by farmers, gardeners, fishermen, and flatworms. Earthworms do not have eyes, but they do have sensors in the top of their "head" that can sense light. This helps them to know when they might encounter conditions that would dry them out. So where do little earthworms come from? And how about the other species? All species must reproduce or they will become extinct. The challenges described above for other issues of moving from the sea to the land also affect reproduction, and significant adaptations were needed and realized. Earthworms are hermaphrodites. That is, both sexes occur in a single individual. However, they still need a mate to reproduce. To do that, two worms align themselves head to toe. The clitellum (the raised band near the head of the worm) forms a slime tube around the work, and the earthworm moves forward out of the slime tube. As it moves along, the slime tube passes over the female pore and picks up eggs. Further along, it collects sperm. In that way, the eggs are fertilized and remain in the tube. The gas exchange that occurs during breathing is facilitated by protein molecules called globins. Globins are heme-containing proteins involved in binding and/or transporting oxygen. (Humans have hemoglobin, which has a similar role.) The hemes bind oxygen. Earthworms absorb oxygen from the atmosphere directly through their skin. Capillaries just under the cuticle absorb oxygen and release carbon dioxide. This might sound inefficient, but it does take advantage of the whole surface area of the worm. Earthworms must be kept moist for gas exchange to occur, and that is a reason that they have the layer of mucus. But too much water can be a problem for worms too. They can drown. After heavy rainstorms, they have to come to the surface.

Mollusca

The mollusks include snails, octopuses, squid, clams, scallops, oysters, and chitons. They have soft bodies and typically a "head" and a "foot," and the body is often covered by an exoskeleton (e.g., snails and clams have shells, and chitons have plates), and they have a small coelom. A mantle covers the head, foot, and viscera and secretes the shell. Slugs and octopuses have lost their shells. The buccal cavity, or mouth, contains the radula. This ribbon of teeth is used for feeding and is anchored in a cartilaginous structure called the odontophore. They use the muscular foot and/or cilia to move over a layer of mucus.

Many mollusks are marine animals, but some are commonly seen in our backyards: snails and slugs (Nekola, 2014). Interestingly, snails have primitive lungs, and some other land mollusks breathe with lungs or gills. A hole in the mantle, the pneumostome, is used for breathing. Those that use gills have an operculum, or trapdoor, that seals off the snail inside its shell to prevent drying out. The head of the snail or slug will have one or two pairs of retractable tentacles. The top pair holds the eyes, which move independently, and the lower pair is for taste and smell. Mollusks use hemocyanin, a copper-based protein, to bind oxygen. Also, whereas human hemoglobin is contained within our red blood cells, the erythrocruorin and hemocyanin are free in the blood of those animals.

The major difference between slugs and snails is that snails have a shell that they can retreat into and slugs have only a small, or no, shell. The shell is made of calcium carbonate. Snails are left- or right-handed (Abe and Kuroda, 2019). To determine which, hold the shell so that the apex (peak) points toward you. If the shell coils in a clockwise manner, the snail is right-handed. If it opens in a counterclockwise manner, it is left-handed. About 90% of snails are right-handed. Many of them are quite beautiful. Collections can be bought on eBay. Slugs and snails tend to feed at night and mostly on decaying organic material, but will also eat fungi, lichens, green leaves, and other small invertebrates. Snails and slugs can cause serious damage to gardens, and some carry disease organisms that affect humans.

Arthropods

By sheer numbers of species and animals, the Arthropoda are the most successful group on Earth. They are also the group that we are most likely to run into in our homes and backyards. Arthropoda includes spiders, crabs, lobsters, insects, mites, scorpions, trilobites, ticks, and many more. They live in every environment in the air, water, and on land. They fall into four or five classes, depending on how they are arranged. Trilobites are completely extinct but are also some of the most recognized fossils. In 1886, archaeologists explored a cave near Arcy-sur-Cure in France that was inhabited 15,000 years ago (AMNH, nd). Among the relics, they found a trilobite fossil from 400 million years ago. A hole drilled in its tail allowed it to be worn by the owner. Chelicerates include spiders, mites, scorpions, and more. Their name is related to the appendages in front of the mouthparts called the chelicerae. In spiders, these are fangs for injecting venom. Myriapods include millipedes, centipedes, and more. Each of their segments sprout one or two pairs of legs or more specialized structures. Crustaceans include lobsters, crabs, shrimp, and crayfish. They are all aquatic except for sowbugs, as noted in Chapter 1. Hexapods include the insects. They are sometimes grouped with the Myriapods to form the Uniramia.

Daddy-longlegs are common around our houses and gardens. There are two types, and both are interesting. Opiliones are typically, but not always, found outside. Although they look like a spider, they are not. They are actually more closely related to scorpions (Giribet and Sharma, 2015). They live under rocks and logs. They have a single body segment and two eyes.

They are not venomous and don't spin webs. Their diet comprises mostly decomposing plant and animal matter, but they will attack small animals, such as aphids or other spiders. They are social animals and sometimes are found in large groups. The second type of daddy-longlegs is a real spider (Huber, 2018). They are most often found inside homes in dark, moist areas. *Pholcidae* have two main body parts, eight eyes, and eight legs. One species is gray, and another has a brown stripe on its ventral side. Like all spiders, they are venomous and spin webs. They eat invertebrates, such as insects and other spiders. So there is a benefit to having them around: they eat bugs in our houses. A long-standing myth has it that these spiders are the most venomous known and could kill a human, but their teeth are too small to deliver a bite. In fact, they are venomous and have fangs, but they are not dangerous to humans at all.

Spiderwebs elicit many reactions. Fear at the sight (or thought) of the spider. Disgust at walking inadvertently into the web. Maybe even awe at the beauty of the web and the skill of the builder. In any case, the web is an amazing structure. Perhaps the most elegant structures are produced by the orb weavers. The cross orb weaver (*Araneus diadematus*) is a common spider throughout the United States and Europe. Their oval webs can be as large as a meter. Liquid silk released from glands at the rear of the spider's abdomen forms fibers as it dries. Amazingly, the glands make six different types of silk, depending on the application needed (Salehi and Scheibel, 2018). A seventh gland secretes the material that makes part of the web sticky. Major ampullate or dragline silk provides the frame of the web. Flagelliform silk forms that capture spiral. Minor ampullate silk forms another spiral that lends structural support to the web. The web is attached to a substrate by pyriform silk. Once an insect stumbles into the web, the sticky aggregate silk holds it in place. The spider quickly bites the insect to inject venom and uses aciniform silk to quickly tie up the prey so that it cannot hurt the spider or the web while the venom works. Finally, cylindriform silk is used to construct an egg case to protect the developing offspring. The silk itself is extraordinary. It is made of protein and rich in the amino acids alanine, serine, and glycine. The resulting fiber is two or three times tougher than Kevlar. Materials scientists are trying to figure out ways to produce it in quantity for various uses.

Spiders are a diverse group, and that diversity is directly related to the evolution of silk (Blackledge et al., 2009). As noted above, the orb weavers use a wet sticky glue to capture the prey. An alternative strategy used by other species uses a different type of dry wooly silk with extremely thin fibers. The cribellate silk fibers use molecular forces called van der Waals interactions and hygroscopic forces to stick to the prey. Beyond the silks, behavioral and other forces added to the diversity.

Insects are the most diverse group on Earth with more than a million species known and maybe 10-fold more yet to be discovered. A statement widely attributed to Jarmila Kukalová-Peck gives some idea of how pervasive insects are: "To a first approximation, every animal is an insect." Insects have six legs and three body segments, and breathe through small openings in their sides called spiracles that are connected to tubes called tracheae. Two types that we often see are ants and bees, and a lot is known about these social insects.

Ants are quite literally everywhere except for the Antarctic. They live in very large nests, often in the ground, and nest members specialize in their labor. They forage for food up to 200 meters from the nest and find their way home by leaving scent trails. They also have sophisticated means of navigating that involve sight, measuring distances, and position of the sun. At times, their wanderings bring them into our homes. Once they find food, long columns appear to carry it back to their nest.

The honey bee we are most familiar with is *Apis mellifera*. The world has 20,000 species of bees, but *A. mellifera* is one of only eight species of honey bees and one of only two domesticated species. Honey bees flit from flower to flower gathering nectar and, in the process, pollinate the flowers. They live in hives and make honey, and they can deliver a powerful sting if aroused. An average hive has 30,000 bees. They produce 10–200 pounds of honey per year. A lot of variables are involved. They forage over a wide area. Beekman and Ratnieks (2000) found that English bees mostly foraged more than 6 km from the hive, and some went as far as 9.5 km. Surprisingly, honey bees are not native to the Americas. They originated in Africa (Honeybee Genome Sequencing Consortium, 2006). The earliest fossils were found in Europe and were about 34 million years old. The Egyptians had domesticated honey bees. They were brought to the New World by early European settlers.

Animals with a Backbone

Members of Chordata have a backbone. Vertebrates are the animals we most often notice. In addition to birds, they include cats, dogs, horses, raccoons, coyotes, skunks, lizards, frogs, and more.

Birds provide a wonderful example of adaptation to specific niches. Darwin himself, in the Galápagos Islands, studied finches, which are not really finches, but types of tanagers. He noted that the approximately 18 species had adapted particular feeding behaviors and that their beaks had also changed to facilitate eating their particular foods. Birding is very rewarding. In most places, it is easy to see perhaps 8–10 different types of birds within an hour or so, even in cities (Lepczyk et al., 2017). Several birds that are commonly found around the United States also demonstrate different sizes, diets, behaviors, and beak shapes and sizes. For example, hummingbirds (family: Trochilidae) have long, needle-like beaks that they used to sip nectar from flowers. American robins (*Turdus migratorius*) eat worms and other small invertebrates in the soil and will also eat berries. Crows (*Corvus brachyrhynchos*) are omnivores. They will eat most anything. Mallards (*Anas platyrhynchoseat*) eat snails, slugs, aquatic vegetation, and grain. Rock doves or common pigeons (*Columba livia*) have gained a bad reputation of eating discarded human food. Canada geese (*Branta canadensis*) eat aquatic vegetation, grass, and grain. The naked heads and powerful beaks of turkey vultures (*Cathartes aura*) may not be pretty, but they do allow them to be important scavengers of carrion.

Fish are vertebrates, and they are cold-blooded. That is, their body temperature is the same as that of the environment. Fish anatomy is distinctive. They use gills to extract oxygen from the water and fins to move themselves through the water. Most are covered with scales and lay eggs. Their streamlined bodies allow them to swim efficiently. A hard, bony plate called the opercula covers the gill inlet on each side of the head. The heart is just inside and behind the gills. The typical fish heart has four chambers, but unlike in mammals, blood flows through the chambers in sequence, then to the lungs, and on to the body. Behind the stomach are fingerlike pouches called pyloric caecae. The number of these is characteristic of the species. Fish also have a swim bladder filled with air that helps them to maintain buoyancy in the water.

Humans are animals too, and we fall into the class Mammalia. The characteristics of mammals include milk-producing mammary glands in females, a neocortex in our brains, hair or fur, and three bones in our middle ear. Like humans, cats, dogs, and squirrels are all mammals. However, the lowly rat is perhaps most like the first mammals (Yuan et al., 2013). Mammals appeared about 300 million years ago when the Earth was dominated by the dinosaurs.

What Distinguishes Humans from Other Animals

We have a lot in common with other animals, especially other mammals. To summarize, we have all three primary cell layers (ectoderm, endoderm, and mesoderm), segments, a coelom, an internal skeleton, and hair, and we are warm-blooded. So what really makes us different? Varki et al. (2008) noted that the differences tend to be relative rather than absolute and suggested several: "brain size, hairless sweaty skin, striding bipedal posture, long-distance running, ability to learn to swim, innate ability to learn languages in childhood, prolonged helplessness of the young, ability to imitate and learn, inter-generational transfer of complex cultures, awareness of self and of the past and future, theory of mind, increased longevity, provisioning by post-menopausal females, difficult childbirth, cerebral cortical asymmetry and so on)" (p. 750).

Ultimately, the differences must reside in our genes. However, our DNA sequences are very similar to those of our primate cousins. We overlap with chimpanzees about 96% (Varki and Altheide, 2005). While the overall sequences of humans and chimpanzees are very similar (within a few percent), there are differences, and those differences must be significant. Interestingly, Pollard et al. (2006) showed that some of the differences are in the noncoding regions and that these sequences seem to be associated with hand and brain development.

As Varki's list above indicates, not all of the differences between humans and the other primates can easily be attributed to gene differences. Our ability to communicate in such sophisticated ways was once thought to be one of those features that was unique to humans, but we now know that other animals communicate in various ways. As Varki et al. noted, the differences are relative, and they will undoubtedly need further research to sort out.

Plants

As of 2016, the US Department of Agriculture estimated that about 36% of the United States was forested (Vogt and Smith, 2016). Even in large, urbanized areas, we are surrounded by plants. We value our yards, gardens, and parks. Many people enjoy gardening or simply having green plants in their offices and homes. But even vacant lots sprout a myriad of weeds that seem sometimes to grow right out of the sidewalk.

Plants are just as diverse as animals, and they have had to solve their own set of evolutionary challenges. Many were similar to those of animals, but plants also have unique problems that come with being immobile and dependent on photosynthesis. All life on Earth began in the seas and stayed there for a long time before venturing onto land. That event is one of the most important in the history of life on Earth, and the plants we see today are the descendants of those plants that invaded the land long, long ago (Niklas and Kurschera, 2010). Those that moved onto land had to face a host of new challenges. It is thought that before plants arrived, the dry land was empty except for some bacteria or fungi that might have colonized it. Rensing (2018) speculates that filamentous algae began to grow above the waterline and eventually came onto land. As long as they were in water, they could take advantage of the many benefits that aquatic plants still do. The new environment brought new challenges, including exposure to drying, sunlight, and temperature changes.

Since that time, a lot has happened. Earth is nothing like it was when life began. The early primitive plants quite literally changed the course of living history on Earth. They changed the atmosphere of the early Earth from a reducing environment (an atmosphere of methane, carbon monoxide, ammonia, and hydrogen sulfide) to an oxidizing environment (an atmosphere with oxygen and carbon dioxide). The early reducing atmosphere might have been important for the origin of life (Follmann and Brownson, 2009; Ferus et al., 2017). However, life as we know it now on Earth was only possible after the atmosphere changed. Second, through the process of photosynthesis, plants capture the energy of the sun and render it to a form that can be used by all other living organisms. They use chlorophyl and other pigments to absorb light energy and store that energy in carbohydrates produced from simple water and carbon dioxide.

As they did with animals, biologists have used those engineering solutions to organize the plant kingdom (Streich and Todd, 2014). They use Linneaus's binominal system, and plants with similar characteristics are organized into species, genus, family, order, and class. Many biologists replaced the term "phylum" with "division" for plants, but they have the same rank. The plant divisions include Marchantiophyta (liverworts), Anthocerotophyta (hornworts), Bryophyta (mosses), Filicophyta (ferns), Sphenophyta (horsetails), Cycadophyta (cycads), Ginkgophyta (ginkgo), Pinophyta (conifers), Gnetophyta (gnetophytes), and Magnoliophyta (angiosperms, flowering plants).

Pond Scum

If we notice the plants growing in bodies of standing water at all, we tend to dismiss them. After all, they appear to be just slimy green stuff growing in mud holes. But they are much more than that. In fact, they are amazing collections of plants and animals. Together, they form a miniature Jurassic Park containing highly integrated sets of organisms with fascinating strategies for life. They often include filamentous algae from *Cladophora*, *Pithophora*, and *Spirogyra* typically serve as homes for many microorganisms and tiny animals (Figure 2.3). They appear as green masses that are free-floating or attached to various surfaces. The filaments have no leaves, roots, stems, or flowers, but they form dense mats. They display many adaptations: some float; others are rooted in the sediments. In temperate zones, they appear in the summer and fall on slow-moving or stagnant waters, such as ponds,

Figure 2.3. Blue-green algae (Cyanobacteria: *Nostoc spongiforme*). Some algae produce toxins that threaten the health and safety of living things that come in direct contact.

Reproduced from Carpenter, K. (2016). Cyanobacteria—*Nostoc spongiforme* (blue green algae). U.S. Geological Survey. https://www.usgs.gov/media/images/cyanobacteria-nostoc-spongiforme-blue-green-algae. Public domain license.

lakes, rivers, and ditches. The algae that make up pond scum may be similar to those that first colonized land and changed the Earth (Donoghue and Paps, 2020).

In addition to the algae, ponds contain lots of other organisms (Kannan and Lenca, 2012). A single drop of pond water might contain thousands of single-cell organisms. Small arthropods are visible to the naked eye (about 3 mm). Copepods are tear-shaped and often transparent with a single large red eye and conspicuous antennae. Water fleas, such as Daphnia, are also transparent, and their beating hearts can be seen with a microscope. Ostracods look like tiny clams. Other inhabitants include water mites (Hydrachnidia), which are colorful (red or orange) and are parasites, particularly of mosquitoes. Water bears (tardigrades) are among the most ubiquitous and hardy of all species. Each of their eight short legs ends with a claw. They eat by using a stylet to pierce plant cells, algae, or small invertebrates and drinking their body fluids. Protozoans include amoebas (Sarcodina), which are blobs of protoplasm that move by the advancement of pseudopodia. Paramecia are one of many ciliates in the pond scum. Hydra are predators and just visible to the naked eye. Some algae include spongomonas (*Spongomona* sp.), euglena (single-cell photosynthetic eukaryotes), *Chlamydomonas* (single-cell flagellated alga), and diatoms. Finally, bacteria may reach millions per millimeter.

Land Plants

Moving onto land was a major achievement for plants as well as animals, but the exact relationships among the major plant groups are still not clear (Qui et al., 2006). Strother and Foster (2021) examined spores that suggest that land plants arose by taking advantage of algal genes. The molecular clock (a method of estimating dates based on mutation rates) indicates that plants began invading the dry land 550–600 million years ago, whereas the fossil record suggests 450 million years ago. The adaptations that made this transformation successful can be seen in the plants in our backyards. These adaptations include transporting water and nutrients, life cycle, reproduction, standing up, and gas exchange, and the various solutions to these problems can be seen in representatives of several common groups.

Like animals, plants had to evolve solutions to a completely new set of engineering challenges when they moved onto land. Also like animals, plants gained new possibilities via favorable mutations that allowed them to survive longer and reproduce more. Those characteristics provide the basis for classifying plants and include vascularization, alternation of generations, and the presence of various features (e.g., roots, leaves, seeds, and flowers).

The first characteristic, vascularization, involves the way plants move water and nutrients throughout the plant. A simple comparison will illustrate the first problem. For mosses, which attain a "height" of a few centimeters at most, water is readily available. However, redwoods (*Sequoia sempervirens*) can grow to 100 meters (330 feet), and raising water from the roots to the crown of those trees is a significant feat. Mosses and redwoods evolved different solutions.

Mosses were among the earliest plants to move onto land. They lack true roots, and they secure themselves to surfaces by thread-like rhizoids. However, the rhizoids do not absorb water or nutrients from the ground like true roots. These short plants hug the ground and simply absorb water directly into the plant cells. Mosses also absorb water and nutrients from what is deposited on the surface of the plant. Using this strategy, mosses cannot grow very tall. However, this seeming limitation has certain advantages. Mosses can dry out almost completely but survive. They can also recreate the plant from any fragment that is separated from the main plant. The bryophytes include the mosses, hornworts, and liverworts.

Redwoods have another strategy. They use a vascular system of specialized cells that carries water and minerals upward (xylem) and the products of photosynthesis downward (phloem). In addition, ferns evolved to contain vascular tissues. These tubular structures run vertically in the roots, stems, and leaves. The vascular tissues can be easily seen by cutting through a stalk of celery. The "fibers" are the vascular tissue. However, the magic ingredient in plant structure is a compound called lignin. This biopolymer is a constituent of plant cell walls and is what actually holds the whole plant upright. It gives wood its hardness and straw its resilience. It is a very common polymer, and makes up one-quarter to one-third of the total dry weight of wood. There are several different types of lignin. Lignin cross-links the cellulose and hemicellulose in plants to form a strong supporting structure. Many parts of the plant cell wall are hydrophobic. Lignin has both hydrophobic and hydrophilic regions. Thus, it is widely used in the formation of xylem because it allows water to flow. The tracheophytes include the ferns, club mosses, cycads, horsetails, gnetophytes, conifers, and flowering plants.

Reproduction is different in plants and animals. Animals produce haploid gametes and lack a multicellular haploid phase. Unlike animals, plants have a reproductive cycle that involves both haploid and diploid phases. This sequence is called "alternation of generations" (Niklas and Kurschera, 2010). For part of the cycle, the plants are in the diploid phase and have two copies of each chromosome. Those forms yield embryos (seeds in more advanced plants) that form the next generation. The next phase involves meiosis, in which the number of chromosomes is reduced to a single copy of each. These cells form spores that later combine to reform the diploid phase, and the cycle continues. For example, in mosses, the dominant phase is the gametophyte. These form the green mats that we recognize as mosses. The sporophyte phase exists only in the fruiting bodies that form each year. In ferns, gymnosperms, and angiosperms, the sporophyte phase is dominant, and the gametophyte phase is sometimes hard to find. The gingko (*Gingko balboa*) is an oddity among the reproducing gametophytes. This living fossil has male and female trees and still reproduces through the use of motile sperm. The gametophyte phase is dominant in mosses, liverworts, and hornworts. The sporophyte phase is dominant in clubmosses, horsetails, ferns, gymnosperms, and angiosperms. When all plants lived in water, reproduction was less of an issue. Those plants released motile sperm that swam to the egg to fertilize it. That strategy is more difficult on dry land. However, old habits die hard, and that method of fertilization remains the rule for mosses and ferns. Other groups have since developed pollen and seeds.

Plants need sunlight to growth, and being able to grow higher enables them to capture more sunlight. This need presents challenges in the water and on land. For example, a long stem or "stipe" enables kelp to grow in deeper water and still reach up to where there is more light for photosynthesis. A holdfast—a mass of root-like structures—anchors the kelp to the substrate but does not absorb nutrients from it. Some kelp have gas sacs to help them float more effectively. For example, the bullwhip kelp (*Nereocystis leutkeana*) has a single long stem or "stipe" that is topped with a kind of a float called a pneumatocyst. Interestingly, the pneumatocyst contains carbon monoxide along with other gases.

Standing up on land involves two steps. The plants have to hold onto the ground to stay in one place, and they have to be able to elevate themselves. Among the earliest plants to move onto land were the mosses (bryophytes). These short plants hug the ground. Without true roots, they secure themselves by thread-like rhizoids, but the rhizoids do not absorb water or nutrients from the ground like true roots. Holding on only solves half of the problem of standing up. The plant must have support structures to provide the mechanical strength to allow plants to grow up into the air. Mosses lack these structures and so tend to grow very close to the ground.

The ferns (pteriphytes) appeared about 400 million years ago in the Devonian, and they "solved" these problems with true roots. Roots represent a huge step in evolution. They allow plants to have a firm grip on the earth and to absorb water and soluble nutrients from the soil. Interestingly, roots also add to the fertility of the soils. They effectively weather the rocks in the earth and, in doing so, improve the character of the soil.

The evolution of roots is still not completely clear. One intriguing clue comes from the fungi. Most plants (over 80%) have a symbiotic relationship with a fungus. The *Glomeromycetes* are mycorrhizal fungi that form structures called arbuscules. These structures penetrate root cells and provide phosphorus, sulfur, nitrogen, and other micronutrients to the plant. Some scientists believe that these structures helped plants move onto the dry land. Interestingly, genes involved in this relationship, called syn genes, are found in most plants.

In the water, plants absorbed CO_2 needed for photosynthesis from the water. But how can they breathe on land? Most land plants, including ferns, conifers, and flowering plants, now have stomata, small openings in the leaves that allow gas exchange. However, mosses have them only on the sporophytes.

Now after that review of the key characteristics of the major plant groups, what kinds of plant features can we easily see? We only need two things: the review we just had and a bit of time to look around us. We typically imagine that plants have leaves, stems, roots, flowers, and seeds. However, the most primitive plants have none of these features, and plant groups add one of more of these standard features at every step of the evolutionary ladder.

Liverworts (Marchantiophyta) are perhaps the most primitive of the land plants (Forrest et al., 2006), and they are easy to overlook. They are small and low to the ground and grow in damp areas. They lack the "normal" features of plants. No leaves. No stems. No roots. No flowers or seeds. They comprise a single layer of cells that are attached to the ground by other elongated single cells called rhizoids. The two forms of liverworts are thalloid and leafy. The thalloid liverworts look like small green pancakes. Leafy liverworts contain leafy looking structures that come in two or three rows, whereas moss leaves form a spiral.

Mosses (Bryophyta) are small, nonvascular flowerless plants. Their dense green mats grow in damp shaded areas. They absorb water and nutrients threw their simple leaves that are typically one cell thick. The diploid sporophyte is the dominant phase. Because of their simple nature, a good fossil record is lacking. However, they were clearly present in the Permian at least 250 million years ago.

Maidenhair ferns (*Adiantum* spp.) are common representatives of the ferns (Filicophyta or Pteridophyta). Maidenhairs include about 250 species. They have dark stems and delicate green complex leaves called fronds. Their "fiddleheads" unfurl to form fronds. Ferns are vascular plants. They lack flowers and seeds and reproduce by spores. Ferns are a good example of the plants that are primarily sporophytes. The diploid sporophyte phase uses meiosis (cell division in which the number of chromosomes is reduced from the normal 2n to n). The haploid spore grows into a gametophyte using mitosis. The gametophyte forms a photosynthetic prothallus. The haploid gametophyte uses mitosis to produce haploid sperm and eggs. An egg is fertilized by a motile sperm and stays attached to the prothallus. The fertilized egg is now a diploid zygote that uses mitosis to produce a new diploid sporophyte (a fern plant), and the cycle begins again.

Horsetails are primitive vascular plants. Like ferns, they reproduce by spores. Spores are produced in a cone-like structure at the top of the stem called a strobilus. Horsetails lack true roots and grow from rhizomes in the soil. They grow as hollow stems. Some have branches with primitive leaves. Today only one genus, containing about 20 species, remains. They appeared in the late Jurassic and, about 100 million years ago, their many species dominated the late Paleozoic forests. They typically grow to 1–5 feet, and are found in wet areas, such as near lakes, rivers, and ponds.

Cycads (Cycadophyta) are unusual plants. They typically look like ferns growing out of a tree stump, and they are sometimes confused with palms. Yet, they are also ancient plants. A number of cycad species are on the brink of extinction. They are vascular and feature naked seeds like gymnosperms. The leaves are so deeply incised that they appear to be complex leaves.

Ginkgos (Ginkgophyta) belong to a single species in this division. These began in the early Permian and are also living fossils. The leaves have a unique fan shape, and the trees are either male or female. They are very common in cities.

Conifers (Pinophyta) are common trees and shrubs, such as pines, spruces, redwoods, hemlocks, cedars, firs, and yews (Neale and Wheeler,

2019). They lack flowers, and their seeds are contained in cones. The number of species is small, but conifers are the dominant plants in many areas in the Northern Hemisphere and are important commercial plants. Most are evergreens. They do not lose their leaves in the fall. They tend to have a single main trunk with branches. Their leaves tend to be needle-shaped.

One of the features most associated with plants is flowers. In fact, most of the plants that surround us are either flowering plants or conifers, and most of the plant food that we consume is from flowering plants. Flowering plants (angiosperms) evolved relatively late, but now they are now the most abundant of all plants, with more than 300,000 species (Christenhusz and Byng, 2016). In fact, their rapid expansion puzzled Charles Darwin, who referred to it as an "abominable mystery." The mystery remains to this day, but some researchers have contributed theories to explain it. Chen et al. (2017) suggested that water lilies may be among the earliest flowering plants and are an excellent model for angiosperm evolution. They urge genomic studies of these beautiful plants. Interestingly, the expansion of the angiosperms had a negative effect on the range of the conifers (Condamine et al., 2020).

Lichens

> The lichen on the rocks is a rude and simple shield which beginning and imperfect Nature suspended there. Still hangs her wrinkled trophy.
>
> —Henry David Thoreau (1849)

Lichens are as common as they are unusual (Figure 2.4). They are easily overlooked, but they are estimated to cover about 6%–8% of the Earth's surface (Asplund and Wardle, 2017). They grow on rocks, tree trunks and limbs, walls, and more, and they have adapted to nearly every environment on Earth, even those where other organisms cannot survive. They come in a wide variety of colors, sizes, and forms. They can be red, yellow, orange, green, gray, or brown. Fruticose lichens look like tiny shrubs. Foliose ones are flat two-dimensional leaves. Crustose ones look like they are painted on the surface. Leprose ones are powdery. Filamentous lichens are stringy like matted hair. The thallus or body of the lichen is made up of the filaments or hyphae of the fungus. They form a complex mesh as the filaments join and separate multiple times.

Figure 2.4. Lichens.
Reproduced from Dillman, K. Lichen habitat. U.S. Forest Service. https://www.fs.usda.gov/wild
flowers/beauty/lichens/habitat.shtml.

Lichens are not plants (Lutzoni and Miadlikowska, 2009). They are a hybrid organism—basically a fungus that is living with an alga or a cyanobacterium (Hawksworth and Grube, 2020). Many species of fungi form lichens, but they are almost always ascomycetes with a very small number of basidiomycetes. The fungi gain carbohydrates made by the algae or cyanobacteria, and the algae or cyanobacteria are protected from the environments and gain moisture and nutrients. The fungi lack roots, so the moisture and minerals come from the air. As another benefit, the cyanobacteria can fix nitrogen from the air.

Fungi and Bacteria

Most of the fungi that we see are the mushrooms on our pizza or the toadstools that seem to appear overnight in cool wet weather. Fungi have many colorful names, including witch's hat, death cap, destroying angel, poison pie, lead poisoner, corpse finder, witches' butter, devil's urn, and dead man's fingers. Several are extremely poisonous, but others are delicious. In fact, there are many kinds of fungi. Some, such as the yeast that helps us to make bread and beer, are single cells. Others include the molds, rusts, and mildews. Still others, such as mushrooms, contain many cells. Fungi lack chlorophyll or other light-absorbing pigments that plants use to capture sunlight for energy. They depend on other organisms for their food. They grow in the soil or on decaying matter. Fungi are important to the health of the soil. They help to retain water in the soil, recycle nutrients, and suppress diseases. Very importantly, they aid in the decomposition of organic material by converting organic material that other microorganisms cannot digest into products that those other organisms need.

What we notice mostly are the mushrooms and toadstools, but these are actually only the "tip of the Iceberg." They are the fruiting bodies or reproductive structures. Mushrooms consist of long threads called hyphae that are only a few micrometers in diameter. A single hypha might have only a few or many thousands of cells. Like plant roots, they help to bind the soil into a stable material to support plants and keep air and water in the soil. Fungi lack roots and so cannot move nutrients and water very far. To overcome this problem, they have evolved to assemble hyphae into rhizomorphs. Rhizomorph cords are collections of hyphae laid down in parallel. They function as "roots" for the fungus and allow it to transport food and water over longer distances. Spores are produced by the gills that are usually on the underside of the mushroom cap.

Mycorrhizal fungi have symbiotic relationships with plants. The hyphae of mycorrhizal fungi form a "cloud" around the roots of a plant. Both organisms benefit. An estimated 85% of all plants have these relationships with fungi. The fungal hyphae engage the plant and especially tree rootlets. In some cases, the hyphae actually penetrate the root cells. The fungal mycelium surrounds the rootlets and helps the plant absorb water and nutrients. The plant provides energy sources, such as sugars, and needed amino acids.

As the mycelium grows outward away from the original mycelium, the fruiting bodies form circles called fairy rings. Fairy rings can become quite large, up to 100 feet in diameter. A fungal mycelium expands more as a ring than as a disc. The mushrooms seen on the surface give an idea of where the mass of the mycelium is underground. In some places, such as athletic fields, the growth can be fairly regular and result in circles that have come to be known as fairy rings. Most of the active growth is at the outer edge of the circle. The growth rate is highly variable even within a single species. Growth averages 10–35 cm per year. Fairy rings are most easily seen when the mushrooms appear. But some species (e.g., *Marasmius oreades*) affect the grass and other plants growing near them. The fungus secretes chemicals that decompose the organic material in the soil for its own use. The released nutrients are also a bonanza for the plants and increase their growth. Just behind the outer edge, however, the mycelium is ravenously consuming the nutrients, and none are left for the plants. The grass is weakened and often dies. On the other side of the mycelia ring, the dead grass provides a wealth of nutrients for the grass, and it enjoys a growth spurt. The result of this activity can often be seen on the surface even without the presence of mushrooms.

Interestingly, fungi can be very large and very old indeed. In fact, they may be the largest single organisms on earth. One in Oregon was estimated to occupy nearly 4 square miles. It would cover 1,665 football fields. The fungus is *Armillaria ostoyae*. To compare it to other large organisms, a large blue whale would be about 110 feet long and weigh 200 tons. The fungus is thought to be between 2,400 and 8,650 years old.

The single word that best describes bacteria is "ubiquitous" (McFall-Ngaia et al., 2013). Bacteria are the most abundant organisms on Earth (if viruses are assumed to be nonliving). Estimates are that 2^{10} bacteria exist (Flemming and Wuertz, 2019). They have exploited every possible environment from the depths of the ocean to the tops of mountains and tropical and frigid areas. They are single-cell prokaryotes. Some are pathological to humans, but most are benign or even valuable to humans. For about the first 3 billion years, bacteria and archaea were the only life on Earth. Bacteria often form biofilms or microbial mats that are from a few micrometers to 50 centimeters thick. These are complex arrangements of cells and extracellular components that are difficult to disrupt. The biofilm that most of us deal with is the plaque on our teeth.

Life Is Everywhere

Living organisms, in all their great diversity, are found all around us. They occur in our houses, our backyards, our gardens and vacant lots. Animals, such as raccoons and squirrels thrive in cities. Insects, spiders, and other small Arthropoda are common. Plants grow out of cracks in the sidewalk. Lichens grow on rocks, fences, and tree branches. Microorganisms cover everything. We walk past them every day. But with a little knowledge, we can begin to see how fantastic nature truly is.

Hopefully, with a bit of knowledge and some awareness, we will notice that splendor and take joy in it. Biology is all around us, and we are part of it for better or worse. But we can also enjoy its beauty and wonder.

3

Life beneath Our Feet

The soil is not a mass of dead debris, merely resulting from the physical and chemical weathering of rocks; it is a more or less homogeneous system which has resulted from the decomposition of plant
and animal remains. It is teeming with life.

—Selman Waksman (1952)

"Common as dirt." If there is anything easy to overlook, it's dirt. It's everywhere. It's not always very pretty. Most of the time, we are just trying to not
track it into the house. But dirt or soil is a lot more than that.

As Waksman wrote, soil is an extremely complex and valuable material.
From their studies of soil microorganisms, he and Albert Schatz discovered
streptomycin, which is very effective against *Mycobacterium tuberculosis*, the
causative agent of tuberculosis. That discovery earned Waksman the 1952
Nobel Prize for Medicine and Physiology. Waksman and his colleagues discovered neomycin, which is typically used to treat skin wounds, but can also
be used to reduce the bacterial load in the intestines. They also discovered
other antibiotics, including actinomycin, clavacin, streptothricin, grisein,
fradicin, candicidin, candidin, and more.

Of course, soil is a lot more than raw material for the development of
antibiotics. Soil is a matrix of minerals and organism, living and dead, along
with some liquids and gases. The minerals are eroded from the rocks in the
area that mix together with organic matter. It provides the anchorage and
nutrients for the plants that collect the energy from the sun that ultimately
passes on to animals and humans. A great amount of life is hidden in the soil,
and several critical processes depend on it.

The natural world is all around us. To see it, all we have to do is walk out
the door, and there it is. Mountains, rivers, trees, rocks, birds, flowers, grass,
squirrels, and clouds are all around and above us. It's almost so familiar that
we forget to notice it. The easiest place to forget to look is straight down. Just

The Biology of Us. Gary C. Howard, Oxford University Press. © Oxford University Press 2024.
DOI: 10.1093/oso/9780197664797.003.0003

under the leaf litter in a shallow layer of soil of, at most, a few feet, a whole living world exists. If we had x-ray vision, like Superman, we could look at that world hidden just under the ground. It would look vaguely familiar, but it might be sort of upside down. Instead of grassy lawns, mats of fungal fibers would cover large areas. The trees and plants that dominate our familiar landscape above would be reflected by their roots as leafless plant structures. Ants, beetles, earthworms, centipedes, millipedes, and other small animals that we often see on the surface would still be wandering about, and water would still flow in rivers. There would be no sunlight, and so the landscape would be continually dark.

The underground world is very different from the one we live in, but our life would not be the same without the processes that take place in that alien environment. In fact, it would not even be possible. Seeds would not germinate, and we would lose most of our food supplies. We would lack the nitrogen that we need to make the biochemicals that are essential for our survival. Other nutrients would be forever locked up in dead matter, and there would be no mechanism to collect them even if they were available.

Here in this chapter, we will try to pull back the soil to peak under nature's carpet so we can appreciate the fascinating and critical life and processes that make up this other world that is just beneath our feet. Only by understanding the wide variety of life underground can we truly understand and fully appreciate the life we can easily see above ground.

Earth under Our Feet

The earth feels solid beneath our feet. We walk on it. We build our houses on it. It's always there. No matter how hard we jump on it or pound it, it's still there. It's easy to take it for granted in the course of our normal activities. Even in California, one of the most seismically unstable regions of the world, where I live, the ground seems stable. Earthquakes are relatively rare events. We joke about them out of false bravado. Most days, we never even think about them. But our "solid" earth is anything but that. Over time, everything has changed. Mountain ranges have risen and fallen. Oceans and seas appeared and disappeared. Continents have moved, collided, and pulled apart again. This isn't just geological ancient history. It continues today, and it has happened everywhere on Earth.

We walk on a thin crust that forms the surface of the Earth. Layers of sedimentary rock were laid down over millennia and penetrated by volcanic columns of igneous rock. The crust also contains rivers of water, pools of oil, deposits of coal and ores, fossils, the remains of recently dead animals and plants. And in the top few meters, a whole world of life exists that makes it possible the life for those on the surface. The bedrock that underlies the soil is porous, some more than others. Granite, for example, is relatively nonporous. Other rocks, such as sandstone and limestone, are much more so. Bedrock can become broken and fractured, increasing the amount of space available for water. Limestone can be dissolved by water. This results in cavities that can fill with water.

Water

Water falls to ground as rain or snow, and these flow on the surfaces in rivers, or percolate down through the soil until they reach layers of impermeable rock. Ground water drains slowly through the soil and porous rock and eventually seeps into rivers, lakes, and oceans. The water collects above the impermeable rock. The ground above this "water table" is slightly wet but not saturated. It contains air and some water and supports plant life. The zone below this is saturated so that water fills the pores between rock particles and the cracks in the rocks. Estimates are that about one-third of all the fresh water on Earth is ground water (Taylor et al., 2013). It collects in aquifers that are often tapped by cities for use. Climate change will put additional pressure on the aquifers (Figure 3.1). It takes thousands of years to recharge those supplies. As they are used more and more to supply growing populations, they will be drawn down. In some areas, subsidence may also occur.

Soil

Soil may be as old as dirt, but actually, much of the soil on Earth is from the Pleistocene (2.5 million to about 11,700 years ago. Soil is the thin layer of material that covers the earth except for areas of bare rock, perpetual frost, deep waters, or glacial ice. Soil is a combination of organic and inorganic material, water, and air (Needelman, 2013). The exact makeup is determined by a host of factors, such as the underlying rock, climate, terrain, and the organisms

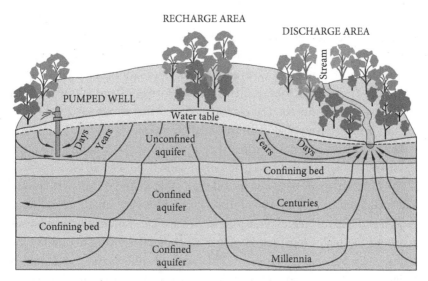

Figure 3.1. Aquifers. Ground water is collected in aquifers that are trapped between layers of impermeable rock.

Reproduced from Winter, T. et al. (1998). Ground Water and Surface Water: A Single Resource. U.S. Geological Survey Circular 1139. https://pubs.usgs.gov/circ/circ1139/pdf/circ1139.pdf.

living there. And it is constantly changing. New material is added as plants and animals die and decay. Wind and rain remove material or deposit new sediments. Water leeches minerals to lower levels of the soil.

A significant portion of the soil either is or was living (Wagg et al., 2014). The soil forms a three-dimensional matrix, and many animals live in that matrix. In fact, they are a key component of the soil environment (Fortuna, 2012). The organic material includes decomposed plant and animal remains that are essentially the same as composted material. Organic matter decays at different rates and with the help of different organisms. Sugars, starches, and simple proteins are broken down easily. Cellulose, lignin, some proteins, fats, waxes, and resins take much longer. The organisms involved include protozoa, bacteria, fungi, earthworms, termites, millipedes, and many more.

The organic material in the soil is called humus, and generally, the more humus, the richer and blacker the soil. Also most of the humus is in the top layer of soil, the topsoil, which is usually a few inches deep. Topsoil is the key to farming. When the first settlers moved into the Great Plains, they found several feet of topsoil. Topsoil takes years to develop, but can be lost quite rapidly. Although the rate of formation of topsoil depends on many factors, most soil scientists believe that it takes more than 100 years to produce 1 inch of topsoil. When settlers first plowed the American Great Plains, they found nearly 2 meters of rich dark topsoil. However, erosion, human development, and poor farming methods have reduced that layer significantly. Thaler et al. (2021; 2022) examined the topsoil loss in the "corn belt," which extends from Indiana westward into the Dakotas. They found that about 33% of the topsoil had been lost from that highly productive region.

The inorganic material is primarily ground-up rocks, resulting from the actions of water, ice, wind, and living organisms. Sand is small pieces of rock (2.00–0.05 mm) made of silicon dioxide and other materials. Silt particles are intermediate sizes (0.002–0.05 mm). Clay particles very small (<0.002 mm) and shaped like tiny plates. Water between the plates forms hydrogen bonds that hold the plates together. However, the bonds are weak and allow the plates to slip past each other easily. This flexibility allows clay to be molded easily. Some of its components have been in the making for millions of years. Others are as new as the leaves that fell last autumn. The organic materials interact with weathered fragments of minerals to form microaggregates of less than 250 microns (Totsche et al., 2018).

Soil is generally laid down as sediments, and so, it usually exists as layers (Figure 3.2 left). The layers can often be seen in ditches or cuts in hills. The layers vary by composition, color, and texture. The colors include brown,

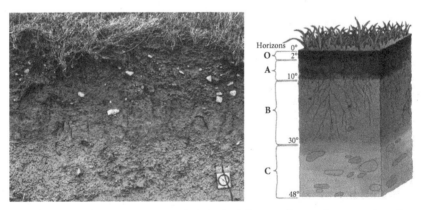

Figure 3.2. Soil horizons. Soil tends to be laid down in layers that are referred to as horizons (left), and those layers are named with letters (right).

Left: Reproduced from Holger, K. (2007). Stagnogley. https://commons.wikimedia.org/wiki/File:Stagnogley.JPG. Public domain license.

Right: Reproduced from Natural Resources Conservation Science. A Soil Profile. USDA. https://www.nrcs.usda.gov/resources/education-and-teaching-materials/a-soil-profile.

yellow, red, gray, white, black, and even green and blue. Scientists refer to the layers as "horizons" (Figure 3.2 right) (Schoeneberger et al., 2012). The horizons are typically lettered. The very top, the "O" horizon, is the leaf litter and other organic matter that covers the ground and is only partially decomposed. Next is the "A" horizon. This is the topsoil. It contains a larger fraction of organic matter and living organisms. The "B" horizon is the sub-soil. The fraction of inorganic material is greater than in the A horizon, and the organic material is less. Other layers follow, each generally with less organic and more inorganic matter. Finally, a layer of solid bedrock is reached.

Life Underground

Many different organisms live just under our feet. We cannot see them; or, we cannot see all of them. We see the parts of plants above the surface, but not the roots. We see the mushrooms, but not the extensive mycelium underground. Plants and animals from single cells to mammals live there. Just underground, a whole different world exists. We do not see it, but it is there nevertheless. Here we will describe just some of the many types of organisms that live in that different world.

Deep Subsurface Life

For some time, living organisms were assumed to inhabit only the upper reaches of the Earth's surface. Life was thought to exist only in the top few meters. However, in recent years, that view has been challenged.

Living organisms have been found at great depths (Gross, 2021), even at 5 km below the surface. For example, a group of methanogenic organisms were discovered deep in the Earth. Amazingly, those organisms have access only to a bare minimum of resources (e.g., basalt, water, and carbon dioxide), but are growing and thriving. They use the basalt as a catalyst to break down water to obtain hydrogen. That hydrogen is then combined with carbon dioxide to form methane. Another organism, *Candidatus Desulforudis audaxviator*, was first discovered in South African gold mines but has since been found in many other areas (e.g., Siberia, California). It obtains energy from natural radioactivity in the rocks around it. Despite its widespread distribution, it seems to have evolved only very slightly since it emerged 55–165 million years ago (Becraft et al., 2021). The environment influences the type and quantity of organisms. For example, some anaerobic organisms use iron-containing minerals (e.g., pyrite), as electron donors for redox reactions. Those areas rich in pyrite, pyrolusite, and siderite feature a greater population than areas that lack ferrous materials.

Those organisms were only the first examples. Since then, many others have been found, and estimates of the number of living organisms in that deep world are many times that of all the humans on Earth combined. They may constitute 10% of the entire biomass of the Earth (Magnabosco et al., 2018).

Soil Macrofauna

The macrofauna are larger organisms, such as earthworms, ants, millipedes, spiders, snakes, lizards, and mammals. Even a few species of birds (e.g., burrowing owls [*Athene cunicularia*]) live underground. Many other animals, large and small, dig into the substrate or use the holes that other animals have dug. These burrows gives them a place to hide from predators, escape the hot sun or the cold of winter, and raise their young. The burrows vary in depth, construction, and complexity. Some are simple tubes of only a few centimeters. Others are complex networks of connecting tunnels with multiple entrances and underground chambers that extend for hundreds of meters. Gophers (family Geomyidae), ground squirrels (family Sciuridae), and groundhogs (*Marmota monax*) are notorious for destroying lawns and gardens. The burrows, large and small, reorganize the soil so that it is aerated and so that water percolates through it more effectively. The tunnels are also used by other smaller organisms that cannot move the earth but benefit by the work of the larger burrowers.

Earthworms (e.g., *Lumbricus terrestris*; there are many other species) are probably the best-known resident in the soil (Medina-Sauza et al., 2019). They slowly dig through the soil, eating decaying matter and allowing oxygen to permeate into the soil. The digested materials that they excrete (i.e., casts, middens) change the soil's composition by recycling nutrients to make it more fertile for plants. The many types of earthworms live at different levels. Some are mostly on the surface in the leaf litter. Some dig tunnels and eat the decaying organic matter that they find in the soil. Still others move organic matter from the surface to underground.

Brown et al. (2000) colorfully described the relationship of earthworms and microorganisms in the soil as the "Sleeping Beauty paradox." Microorganisms play Sleeping Beauty. They rely on organic matter for food, but they are too small to transport that material. Earthworms have the role of Prince Charming. The Kiss is the mucus that is secreted by the earthworms. It contains glycoproteins that "awaken" the microorganisms. Also as the earthworms move through the soil, they transport microorganism to the food and provide them with warmth, moisture, and food. In this scenario, the microorganisms repay the earthworms by providing digested material that the worms can eat.

Ants are one of the most common underground animals. Nearly 22,000 species are estimated to exist. They are social insects and often live in large underground colonies. Ants are everywhere. The only places that lack ants are Antarctica and a few islands (e.g., Greenland, Iceland, Hawaii). Ants are important for the soil. By digging their nests and tunnels, they turn over the soil and aerate it. This allows water and oxygen to reach roots deep in the soil. Ant refuse materials are a source of nutrients for the soil (Farji-Brener and Werekraut, 2017). Soil fertility is greater in nest areas, and the refuse material was richer in dry habitats than grasslands. Plants grow more vigorously on and have more roots in the soils of ant nests than in adjacent areas. The ants also carry seeds into their nests to eat, but many of the seeds germinate to start new plants.

Soil Mesofauna

Mesofauna are invertebrates of 0.1–2.0 mm and include earthworms, ants, nematodes, mites, springtails, proturans, pauropods, rotifers, tardigrades, spiders, pseudoscorpions, harvestmen, isopods, and millipedes and centipedes. There is some overlap with the megafauna (e.g., earthworms). Springtails are important in recycling organic material. Their small fecal pellets are ideal food for microorganisms. Densities of springtails can reach 10–100,000/meter^2. Ground beetles are both predators and herbivores. They eat aphids, snails, and springtails. These organisms feed on a variety of organic material, both living and dead. They also eat fungi, algae, lichens, and other animals. They also clean up decaying and fecal matter. Mesofauna live in the pore spaces between the soil particles. Nematodes and other animals feed on bacteria and other material.

Bacteria

Bacteria are the most numerous of all living organisms (Delgado-Baquerizo et al., 2018). A gram of soil may contain up to a billion bacteria. They live in the tiny amount of water held between the soil particles. Most require oxygen. Those that do not tend to be involved in putrefaction, a critical function. Many types and sizes of animals live in the soil, and they are a key component of the soil environment. They help to break down plant and animal matter so it can be recycled. Soil bacteria perform important functions. Some turn atmospheric nitrogen (N_2) into ammonium. Others change the ammonium to nitrite (NO_2-) and nitrate (NO_3-). One group of those soil bacteria, the actinomycetes, give soil its "earthy" smell and are involved in decomposition that enriches the soil.

Fungi

The fungi are also common (Frąc et al., 2018). A gram of soil might also contain a million yeast or molds. We might ordinarily think of fungi as the mushrooms on our pizza or the toadstools that seem to appear in our backyard overnight in cool wet weather. The fungi are both microscopic and enormous. Some are parasitic. Many live with plants in a close and mutually beneficial relationship. The fungi break down dead organic matter and provide nutrients to the plants. The plants, in turn, supply nutrients to the fungi.

Usually, only the fruiting bodies of the fungi, the mushrooms, are visible. These are actually only the "tip of the iceberg." They are the fruiting bodies or reproductive structures that produce spores. Spores are produced by the gills that are usually on the underside of the mushroom cap. Some mushroom caps are delicious, but others are extremely poisonous, and many poisonous mushrooms look very much like safe ones. Never eat any mushroom unless it has been identified as edible by an expert. The main body of the fungus is underground and connected by a maze of thin filaments called mycelium, and the total organism can be extremely large and long-lived. A specimen of *Armillaria ostoyae* discovered in Oregon's Blue Mountains is thought to be the largest living organism on Earth (Casselman, 2007). It covers 2,384 acres (10 square kilometers). Based on its growth rate, its age is estimated to be between 2,400 and 8,650 years.

Fungi are important to the health of the soil. They help to retain water in the soil, recycle nutrients, and suppress diseases. Very importantly, they aid in the decomposition of organic material by converting organic material that other microorganism cannot digest into products that those other organisms need. Like plant roots, they help to bind the soil into a stable material to support plants and keep air and water in the soil.

Roots

A major portion of life underground involves roots. Trees dominate the landscape, and their roots make up much of the living mass underground. Amazingly, the roots are often even larger than the tree above the ground. Because the roots are underground and out of sight, it can be difficult to appreciate just how much ground they encompass. In fact, the rhizosphere is a crowded space. Roots, root hairs, mycelia, rhizomes, tubers, rocks, living organisms, and more are packed together. They compete and cooperate for space, water, nutrients, and oxygen.

Vascular plants (e.g., ferns, conifers, and flowering plants) have roots that provide structures that support the plant and channel water up into the rest of the plant. They provide stability to the soil and help to prevent erosion by giving the soil structure. Mosses and mushrooms have rhizomes, underground stems that grown laterally. Grasses have stolons. Roots, tubers, and some nuts from underground provide valuable food sources for humans and other animals. Potatoes, carrots, onions, sweet potatoes, yaro, and many others store starches for the plant and also for human consumption.

Through their roots, plants have extensive relationships with fungi. For some time, it was thought that carbohydrates were the main contribution of plants to the fungi. However, lipids or fats are also exchanged. Rich et al. (2021) examined the genes involved in lipid synthesis in the primitive plant *Marchantia paleacea* and found that the arrangement evolved very long ago.

The rhizosphere is loosely divided into three zones (McNear, 2013). The endorhizosphere comprises parts of the cortex and endodermis. In this region, microorganisms can live between the root cells. The rhizoplane is an intermediate zone that includes the epidermis and mucilage. The ectorhizosphere is the outer zone. These three zones are only general terms. Great variations exist from species to species.

The growing tip of the root is covered with a root cap that helps protect the root. The root exerts very large amounts of pressure on the soil as it pushes its way along. The root meristem, the growing region, is just behind the root cap. The root cap and root meristem are programmed to lose cells as they are worn out and then replaced by new cells from the meristem.

The root also secretes various materials (e.g., mucilage, organic acids, amino acids, proteins, sugar, phenolics) as it grows. Estimates are that roots release 10%–40% of the carbon that they fix in photosynthesis. Much of that material is useful to symbionts, such as fungi, living in the soil. The secretions also liberate minerals and nutrients from the soil so they can be absorbed by the roots.

Key Processes Underground

Seed Germination

Plants are essential to the survival of nearly all life on Earth. The energy of sunlight is captured by photosynthesis, and land plants provide food for humans and many other animals. So seed germination is a critical part of growing plants, and germination for the conifers and flowering plants normally takes place underground. Seeds are the next generation of plants. Their germination is an almost magical process in which the plant embryo "wakes up" and begins an orderly series of steps to produce a new plant. Non-seed plants, such as ferns and mosses, also germinate from spores, and the basics are the similar (Setlow, 2003). Spores can also survive inhospitable conditions. When the conditions are right, the spore opens to allow the organism to grow.

Germination is the beginning of the growth of a plant from a seed or a spore (Weitbrecht et al., 2011; Rajjou et al., 2012). Seeds are remarkable things. They are the embryos of plants, and they are among the latest evolutionary developments. Seeds only occur in gymnosperms (conifers) and angiosperms (flowering plants). They are lacking in ferns, mosses, liverworts, horsetails, and other lower plants. The seed contains the plant embryo and a food supply of starch, protein, and oils for the developing plant, called the endosperm. In angiosperms, all of this is all packed inside a tough seed coat that forms from the maternal tissues of the ovule. Interestingly, fertilization involves two sperm and one egg. The zygote remains fairly quiet for some time, but the endosperm develops rapidly. Seeds remain dormant until the environmental conditions are conducive for the growing plant. Those conditions are a specific combination of temperature, water, oxygen and light. Even then, a portion of the seeds do not germinate. The belief is that this is a strategy that plants use to survive a late frost or other unusual circumstance. Not all seeds sprout. They can be damaged by animals or weather. Since they are living organisms, they need to germinate within some period or the seed dies. In germination, the metabolic machinery of the seed is reactivated from its dormant state. The first parts of the plant that begin to grow are the radicle and the plumule. The radicle forms the root, and the shoot forms that will emerge from the soil.

Germination depends on several factors. For example, water is critical. Water is taken up by the seed, and the seed swells so that the seed coat ruptures. The radicle and shoot begin to grow and expand. Once the first leaves or cotyledons break through the surface of the soil, germination is complete, the endosperm is used up, and photosynthesis begins. Sulfur metabolism is another key factor (Mondal et al., 2022). It affects sulfur-containing storage proteins, and sulfur transporters. Sulfate is taken up by plant roots and dispersed throughout the plant. Cysteine acts as the primary donor of sulfur to other organic compounds in the plant.

Dormancy and germination are distinct processes. Each is critical and controlled by hormones. The transition from dormancy to germination occurs under the appropriate conditions (Shu et al., 2016). Dormancy is regulated by the opposing actions of abscisic acid and gibberellins. Auxin and the AP2-domain-containing transcription factors are also involved. The balance among these hormones, in addition to the environmental conditions, determine when dormancy ends and germination begins.

Nitrogen Fixation

> Water, water, every where,
> Nor any drop to drink.
> —Samuel Taylor Coleridge, "The Rime of the Ancient Mariner"

Coleridge could have said the same thing about nitrogen. The air around us is about 78% nitrogen. The pure elemental nitrogen in the air is a gas consisting of two atoms of nitrogen bound together. It's the seventh most common element in the solar system and arrived on Earth in the chunks of material that collided to make up the primitive Earth. We are quite literally surrounded by it, and we breathe it in with every breath. There is plenty of nitrogen around, and that's a good thing because it's a critical element for life. We and all life on Earth could not live without it. Yet, amazingly, the nitrogen all around us is completely unavailable to us.

Nitrogen is part of many key biological molecules. A few examples will demonstrate the importance of this element. Green plants use chlorophyl for photosynthesis so that the energy of sunlight can be stored in carbohydrates. The cellular currency of energy is ATP and NAD molecules, and both contain multiple nitrogen atoms. Proteins are key molecules for all life forms. They are used to build enzymes, structural elements, and antibodies, and each amino acid that makes up a protein contains at least one molecule of nitrogen. Finally, the nucleic acids that form our genes and carry our genetic information include many nitrogen atoms. Clearly, life as we know it depends on nitrogen.

So, if nitrogen is important and we can't obtain nitrogen by breathing it in, how do we get it? We get that nitrogen from the foods we eat. So do animals. However, the nitrogen in our foods is not elemental nitrogen. It exists in various chemical combinations that can be metabolized and reused over and over again. As plants and animals decompose after death, their nitrogen-containing compounds are released into the environment for plants to absorb and use. Some of the nitrogen is also released as volatile forms (e.g., nitrogen oxides, ammonia) and returns to the atmosphere. Some nitrogen is reduced in the atmosphere by lightning. In more recent years, humans have learned how to make fertilizer that contains artificially reduced nitrogen (the Haber-Bosch process). However, the most significant source of reduced nitrogen is nitrogen fixation.

Those are not the major sources of nitrogen compounds. Plants and animals cannot fix nitrogen, but fortunately, many species of bacteria and archaea can do just that (Mylona et al., 1995). Those organisms, known collectively as diazotrophs, include the green sulphur bacteria, Firmibacteria, actinomycetes, cyanobacteria, and all subdivisions of the Proteobacteria and some Archaea (single-celled organisms that are distinct from bacteria and eucaryotes) (Dixon and Kahn, 2004).

The process of biological nitrogen fixation was first described by the Dutch microbiologist Martinus W. Beijerinck (1901). It is a complex chemical reaction and an energy-intensive process for the plant (Figure 3.3).

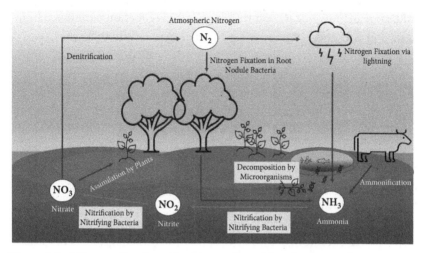

Figure 3.3. Nitrogen fixation. A basic overview of the nitrogen cycle. Red arrows represent the parts of the nitrogen cycle (decomposition and nitrification).

Reproduced from the Columbia Environmental Research Center. (2021). Basic overview of the nitrogen cycle. U.S. Geological Survey. https://www.usgs.gov/media/images/basic-overview-nitrogen-cycle. Public domain license.

Nitrogenases catalyze the chemical reduction of dinitrogen to ammonia, which is easily taken up by plants and used to synthesize various nitrogen-containing compounds. Aspects of the structure and function of these metalloenzymes have been conserved throughout evolution. They have two components: a smaller iron protein and a larger molybdenum–iron protein. Some also have a nitrogenase that contains vanadium-iron or iron-iron. The reaction involves breaking the three covalent bonds that hold the two nitrogen atoms together. Three hydrogen atoms are added to each nitrogen atom to form two molecules of ammonia. The enzymes overcome the need for the enormous energy ordinarily needed to do this. Even so, the reaction requires 16 moles of ATP to reduce each mole of nitrogen. That energy comes from several sources, depending on the organisms that are doing the fixation. Some microorganisms, such as the cyanobacteria, use photosynthesis to obtain the energy for the reaction. Others depend on symbiotic relationships with plants in which the bacteria exchange ammonia for carbohydrates from the plants. The exchanges take place in plant rhizomes. Finally, some heterotrophic soil bacteria (e.g., *Azotobacter, Bacillus, Clostridium,* and *Klebsiella*) obtain energy by using organic molecules from decomposing matter or by utilization of inorganic molecules. The nitrogen reduction is inhibited by oxygen, and thus, these organisms are completely or mostly anaerobes.

The critical reaction involves a protein called leghemoglobin that contains a complex molecule called a heme group with an iron atom in the middle. Nature often reuses the same basic designs for proteins and enzymes, and heme groups are also found in many other proteins, including hemoglobin, which carries oxygen in our own blood. With its heme group, leghemoglobin controls the level of oxygen so that another enzyme nitrogenase can use nitrogen to produce ammonia. Of these methods, the most important is symbiotic nitrogen fixation (Wagner, 2011), and the most important pairings are between *Rhizobium* and *Bradyrhizobium* bacteria and legumes, such as alfalfa, beans, clover, cowpeas, lupines, peanut, soybean, and vetches. Rhizobium forms up to 1,000 small growths or nodules on the roots of some plants. The bacteria attached to the root hairs through a calcium-binding protein called rhicadhesis. They then secrete Nod factors that cause the root hairs to curl up into a "shepherd's crook." The bacteria can then penetrate into the root hairs and begin to form the nodule. They lose their cell walls and form branching cells called bacteroids. The bacteria gain food from the plant, and the plant benefits greatly from the nitrogen that is fixed.

Some plants are better than others at fixing nitrogen. Peanuts, cowpeas, soybeans, and fava beans do a good job, but alfalfa, sweet clover, true clovers, and vetches are among the best. Farmers plow these crops under so that other plants can then absorb the nitrogen compounds that are released from the decaying plants.

Plant Interactions

Plant roots interact with the fungi in the soil (Bais et al., 2004). In a sense, those interactions are a form of communication. Exudates containing amino acids from the roots attract bacteria and fungi to the roots to establish a symbiotic relationship in a process called chemotaxis. Moreover, some release compounds also regulate genes in the other symbiotic partner. For example, isoflavonoids and flavonoids in the exudates of some legumes activate the *Rhizobium* genes that control the formation of nodules for nitrogen fixation. In some roots, ion transport at the root surface results in electrical fields that attract (electrotaxis) other organisms (van West et al., 2002). *Phytophthora palmivora* and *Pythium aphanidermatum* caused the Great Potato Blight in Ireland. Their zoospores are attracted by electrical fields.

Plants send and receive signals to report on their environment. Sharifi and Ryu (2021) described these as "wired" and "wireless" communications. The wired communications involve signals received and sent via the mycorrhizal hyphae and parasitic plant stems. A good example is dodder (*Cuscuta* spp.). The wireless communications involve molecular signals that diffuse through the soil. In either case, the signal causes the recipient plant to adapt to the changing environment. Signals can influence neighboring plants with regard to lateral root formation and seed germination. Domestication has reduced the ability of some plant crops to communicate effectively, and this affects their yield.

Decomposition: The Ultimate in Recycling

All living things eventually die, and their death begins the process of decomposition. Many of the processes in plants and animals are similar, and that breakdown is critical to recycle nutrients to feed new plants that we ultimately depend on. The molecules and atoms that make us up are only on loan to us. Eventually, we have to give them back so other living things can use them.

Decomposition is one part of the carbon cycle (Battin et al., 2009). Carbon is a major component of all living organisms. It cycles from the atmosphere to living organisms and back again after their death. Only a very small part (about 0.04%) of the air we breathe is carbon dioxide. Each breath that we exhale is about 4% CO_2. Much of the Earth's carbon is sequestered in marine sediments and limestone deposits. Those are the skeletal remains of animals that have been laid down for millions of years. Fossil fuels, such as coal, oil, and natural gas, also store carbon.

Another part of the carbon cycle involves the processes of living plants (Janzen, 2004). For example, they use photosynthesis to capture the energy of the sun. They capture carbon dioxide from the air and convert it into sugars and starches that later form energy sources for animals and humans. Plants also synthesize vitamins and other important nutrients or otherwise make them available.

As with animal decomposition, an ecosystem forms when plants begin to decompose. That ecosystem contains organisms that feed on various parts of the plant. They all take their turn eating as different molecules are exposed in different parts of the process. Primary consumers, including bacteria, fungi, nematodes, some types of mites, snails, slugs, earthworms, millipedes, and sowbugs, eat the organic matter. Secondary consumers, such as springtails, other types of mites, some types of beetles, nematodes, rotifers, and soil flatworms, eat the primary consumers. Tertiary consumers, such as centipedes, ants, spiders, and other types of beetles, eat the primary consumers. The higher-level consumers contribute excrement and their own dead bodies to the organic material to begin the process again. Earthworms participate in decomposition in two ways. They eat dead plant material and leave their droppings. In addition, the holes they dig allow oxygen and water to penetrate the decaying material, giving other small decomposers more access to the material and encouraging chemical reactions. Also in soil, the decomposed cell walls of plants remain and help bind together clay particles and other elements of soil. This gives the soil pores so that air and water can penetrate into the ground, and it gives a structure for the organic components to stick to so they are not washed away.

So what exactly happens during plant decomposition? While many of the degradation processes are similar for plants and animals, there is one significant difference. The cells of both plants and animals have a cell membrane, but plant cells also have a tough cell wall made of cellulose, hemicelluloses, and lignin. The first step in plant decomposition is breaking the cell wall. Of course, in biology, where there is a need, there is a solution. Many species have evolved to take part in the breakdown, including invertebrates and insects in the soil and fungi. These materials of the cell wall decompose mostly by biochemical methods and at different rates, depending on temperature, soil conditions, and water. Lignin, for example, resists most organisms, except some fungi, such as white rot fungus.

The transformation of organic matter is a fascinating process. Fungi and bacteria start the process. They have enzymes that break down cellulose and other plant components (enzymes that many small animals lack). The biochemical compounds that make up plants are extremely complex, and the organisms involved in decomposition use a variety of enzymatic reactions to decompose plant material. Oxidation is an important reaction. This is the same basic reaction that occurs when wood burns, but the reaction is much slower. A plant carbohydrate, protein, or lipid combines with oxygen to produce carbon dioxide and water. In biochemistry, the carbohydrate or other compound is oxidized and the oxygen is reduced. Thus, these reactions are called redox (for reduction and oxidation) reactions.

Other enzymes attack specific molecules. Chitinases catalyze, or speed up, the degradation of chitin. They have been found in many organisms, such as bacteria, fungi, plants, invertebrates, and vertebrates. Endochitinases cut chitin inside the molecule, and exochitinases degrade chitin from the ends. Cellulases break down cellulose, a complex carbohydrate. Various proteases digest proteins either by working from the ends of the protein molecule or by cutting in the middle.

The most important microorganisms are bacteria, and they make up about 80%–90% of the microorganisms in the soil and in decomposing matter. Actinomycetes give soil its "earthy" smell. In decaying material, they look like spiderwebs with long gray filaments. They degrade complex organics (cellulose, lignin, chitin, and proteins). Fungi (molds and yeasts), which tend to live on the outside of the decay, also break down complex plant components and, in that way, help bacteria to do their work. Protozoa (one-celled animals), rotifers, and other microscopic animals that live in water droplets feed on the organic material, bacteria, and fungi.

The decomposition reactions are greatly affected by the reaction conditions. As just noted, oxygen is used in the reactions, and carbon dioxide is produced. If all of the oxygen is used up, a different set of reactions will take over. These use sulfur and produce a smell like rotten eggs. Acidity is also critical. The pH scale, which varies from 1 to 14, is used to measure acidity. A relatively neutral pH of 5.5–8.5 is ideal for decomposition. The reactions themselves release organic acids, which early on, favors the growth of fungi.

Composting is a process of aerobic decomposition of plant matter that many people are familiar with. Many kinds of microorganisms are involved, and each prefers a particular type of organic material. Composting also depends on physical conditions, such as temperature, moisture, and aeration. Decomposition reactions release heat, and the temperature of a compost pile can reach 40°C–50°C (104°F–122°F) within a few days. The heat encourages growth of microorganisms that are involved in the decomposition. But there is a limit: if the pile gets too hot, the microorganisms will die. This is why the pile must be turned periodically. It aerates the pile and ensures that the reactions continue optimally. The amount of water in the pile affects the temperature and the reactions, as well, as water tends to hold heat. If the plant material has been broken down too much, the small particles can become compacted and block the circulation of oxygen, which will slow the reactions.

Sometimes plants (like animals) do not decompose, becoming fossils. The process of fossilization can involve either permineralization or replacement. In permineralization, or petrification, water containing large amounts of calcium carbonate or silica is absorbed by relatively porous organic plant material. Then, time and pressure stabilize the minerals and preserve the plant structures. In replacement, water dissolves the plant material, and the remaining space is filled with the minerals. The result in either case is a fossil replica of the original living tissue.

Plants can be preserved over the long term in a different manner as well. In the Carboniferous period (300–350 million years ago), tree ferns (*Filicales*), horsetails (*Equisetales*), club mosses (*Lycopodiales*), and other plants covered much of the earth in swampy forests. When those plants died, the water and mud of the swamps hindered normal degradation. Some of the dead organic material decayed under anaerobic conditions (in which little or no oxygen was present), forming a thick layer of peat. In some cases, millions of years' worth of heat and pressure from overlying sediment squeeze the water out of peat, changing into it coal. Coal, sometimes called "fossilized sunlight," is more than 50% carbon by weight. There are several types of coal, including lignite, sub-bituminous coal, bituminous coal, anthracite coal, and graphite (pure carbon). Ten feet of peat produce about one foot of coal.

Like coal, petroleum also comes from living sources. Zooplankton and algae died and settled to the bottom of a sea or lake. They later were covered by mud and other sediments. After millions of years of heat and pressure, they changed into various compounds. For example, kerogen is organic material found in oil shale. Kerogen contains carbon disulfide or bitumen, which the ancient Egyptians used to prepare mummies of their dead. It is also useful for waterproofing boats. More heat and pressure turn kerogen into oil and natural gas.

Think about What's Underground

The next time that you walk across some open ground take a moment to ponder what is going on under your feet. The soil provides a home for many organisms, including worms, insects, bacteria, and fungi. Many of these have a function that benefits us, such as decomposing dead organic material, aerating the soil, and more. The soil also provides nutrients to the plants that capture the energy of the sun and that ultimately provides energy to us. In addition, we get nitrogen from reactions in primitive organisms that fix the nitrogen from the air so that we can use to make compounds critical for life. The soil bacteria are essentially never seen by us. Plants begin their growth by germinating from seeds underground. And ultimately, all living things die and begin to decompose so that their chemical constituents are released and recycled to feed other plants. We borrow our molecules and atoms for a while. In the end, we give them back so they can go to other living things. Most of these events occur underground where we cannot see them.

4

Supermarket Systematics

Every moving thing that liveth shall be meat for you; even as the
green herb have I given you all things.

—Genesis 9:3, King James Version

Food is critical for life. It provides the energy and raw materials for our existence. Our hunter-gatherer ancestors spent much of their time seeking food
(Dyble et al., 2019). They gathered edible plants, berries, nuts, and roots.
They hunted game, and they likely scavenged as well. They needed a considerable area for their activities. Fixed camps were not possible, and their diet
changed with the seasons as different plants and animals became available.

The transformation from hunter-gatherers to agriculture about
13,000 years ago is one of the landmark achievements of humans (Diamond,
2002). For early humans, it was a lot harder to farm than to hunt and gather.
However, the benefits, and the farmers, rapidly accrued. The farming revolution began most prominently in two widely separated areas (i.e., the
Middle East and China), along with seven others (i.e., Mesoamerica, Andes/
Amazonia, eastern United States, Sahel, tropical West Africa, Ethiopia,
and New Guinea). Yet most of what we eat today originated in the Fertile
Crescent. Some of it was serendipity. For example, a single-gene mutation prevented the seed pods from releasing grain so that wheat and barley
could be easily harvested by early farmers. Certain animals were domesticated for specific purposes. Some for food. Some for fiber. Others for protection (e.g., dogs). One intriguing aspect is that humans domesticated so few
plants and animals of the many that could have been selected. Good possible
explanations exist. Some species were easier to domesticate or had greater
value than others. Some were readily available in the wild and did not need to
be domesticated. Still others grew too slowly, were dangerous, or would not
breed in captivity.

The Biology of Us. Gary C. Howard, Oxford University Press. © Oxford University Press 2024.
DOI: 10.1093/oso/9780197664797.003.0004

Today we hunt for bargains at the local supermarket and gather what we need into a shopping cart. Supermarkets offer a wide selection of foods, and even the seasonality of different plants and animals has been lost as foods are transported even from hemisphere to hemisphere regardless of the season.

Ultimately, our food comes from living or previously living plants and animals. We enjoy the meats of different animals, fish, and shellfish. Lots of beautiful vegetables and fruits, along with nuts, mushrooms, and grains, are stocked in the produce section. Eggs, milk, cereals, and many other foods adorn the shelves. Interestingly, a biologist might view a supermarket as about as good for learning about living organisms as our backyards or even a zoo. The difference is that, at the supermarket, we can look at things close-up as we want and even take them apart. In this chapter, we will look at the living organisms in another corner of our environment: the supermarket. These animals and plants complement those that we reviewed in Chapter 2. Of course, we also rely on these same plants and animals for food, and we will look at the process of eating and digestion in Chapter 8.

So let's take a walk around a supermarket to see what we can learn. A supermarket isn't a zoo or museum, but it does have one great advantage over those other institutions. We can't touch the exhibits in a museum or touch the animals in the zoo. But we can buy the plants and animals in the supermarket and take them home to examine more closely. In this section, we will do just that.

Animals

Alert

Biology is fascinating, but it isn't always pretty. It can be messy, even disgusting. And sometimes it involves looking carefully at things that aren't pleasant. This is especially true of animal food products. Most of us only deal with meat when it is served to us for a meal. Few of us now live on farms and ranches or work at slaughterhouses. The processing of animals for meat is hidden from us. The meat counter reveals a lot about various tissues. Just looking at them, we can see a lot about their anatomy and physiology, and much of that is similar to human anatomy and physiology. If you might be bothered by this aspect, please skip this section on animals and move on to the sections on plants and fungi.

Meat

Meat has been a key component of the human diet for millennia (Domínguez-Rodrigo et al., 2014). Unlike our early human ancestors, most of us don't actually kill animals for food. Few of us have visited a slaughterhouse or even seen a butcher at work (Lawrie, 2006). All of that takes place out of sight. We see neat packages in the counter. On some level, we know that those cuts of meat were once living animals, but we are insulated from that knowledge by the protective cellophane wrappers.

Even though we call ourselves omnivores, we are quite picky about the animals we use for food (see Chapter 2 for a review of the animal phyla and the plant divisions). With few exceptions, we only eat animals from three phyla: mollusks (clams, oysters, squid), arthropods (crabs, lobsters), and chordates or vertebrates (fish, birds, mammals). In fact, animals with backbones form the main part of our meat diet. A typical supermarket usually keeps several types of fish. Birds include chickens, turkeys, and ducks. Beef (cow), lamb or mutton (sheep), and pork (pig) are common. Sometimes, more exotic animals, such as rabbit and buffalo, are available. Even wild game that we might occasionally encounter includes only vertebrates, such as elk, venison (deer), bear, boar, and geese.

Meat is mostly muscle tissue, and the meat counter contains various cuts of muscle from cows, pigs, chickens, and other animals. Those animals and we ourselves have three types of muscle (Noto et al., 2021). The heart is made of cardiac muscle, and smooth muscle lines arteries and uterus (Hafen and Burns, 2021). But most of the meat in the butcher's counter is striated muscle. This type forms the muscle mass of our bodies and that of other animals. We and animals use muscle tissue for movement, and so, other structural elements (e.g., bone, connective tissue, cartilage, and blood vessels) that allow the muscles to work are closely associated with the muscle.

The most obvious feature of meat is its color, red or white. The red or white colors reflect the functions of the muscle. Red meat, such as beef, contains "slow-twitch" muscle. The red color comes from the iron-rich and heavily pigmented protein myoglobin that stores a ready supply of oxygen to produce energy aerobically (with oxygen) for continuous activity, such as standing or walking. White meat, such as in fish, contains "fast-twitch" muscle, which is used for quick bursts of activity, such as escaping a predator. These muscles use glycogen anaerobically (without oxygen) for energy. Chicken contains both white and dark meat. The bright red color of beef comes from the oxygenation of the myoglobin (Suman and Joseph, 2013). If the meat is vacuum packaged, it will be darker, but will become bright red once the packaging is broken and the meat is again exposed to oxygen.

Muscle also shows a fibrous "grain." Those striations are bundles of muscle fibers. The main proteins that make up the muscle fibers, actin and myosin, are polymers of amino acids that overlap with each other sort of like interlacing your fingers (Sweeney and Hammers, 2018). The muscle contracts when these two proteins "ratchet" past one another (Figure 4.1). Muscles generally work as pairs in opposition to each other. They can pull but they can't push. When one muscle contracts, the opposing muscle relaxes, and its fibers go back to their original positions. The fibers can be easily separated from each other, but they resist stretching. The grain in filet mignon is small, and the meat is tender. It is easy to cut the meet in any direction. In cuts from muscles that work much harder, such as flank steak, the grain is more defined. The meat is better if it is cut "against the grain" to shorten the muscle fibers.

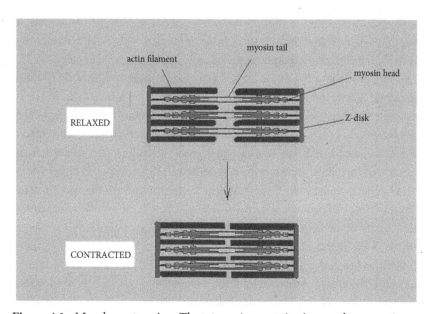

Figure 4.1. Muscle contraction. The two major proteins in muscles are actin and myosin. They are interlaced, and on contraction, they rachet pass one another.

Reproduced from Tinastella. (2008). Myosin actin binding. https://commons.wikimedia.org/wiki/File:MYOSIN_ACTIN_BINDING_CHEM114A.jpg. Public domain license.

Meat involves more than muscle tissue. Animals store the energy from their food mostly in the form of fat. As noted earlier, fat consists of triglycerides, which are a combination of glycerol with one to three fatty acid molecules. Fat is a source of energy for the animal, but for us, it is also a source of flavor (Wood et al., 2008). Fat makes the meat more tender because it melts as the meat cooks and tends to keep the meat moist. A cut of meat is referred to as well "marbled" if it has significant and visible amounts of fat throughout. Of course, now we know that animal fat also has health risks, and health authorities advise us to temper our consumption.

Other structures can be seen in meat, including those that form the support structure for the animal: the bones and cartilage. The bones are the main parts of the endoskeleton that is characteristic of vertebrates (Clarke, 2008). Bones are made of calcium and phosphorus, which form a strong structure for support and provide attachment points for muscles to pull against a solid structure so they can contract. But bones, particularly the larger bones, are not just a solid structure. A solid outer layer, the cortical bone, surrounds an inner honeycomb structure. The inner trabecular bone contains the marrow that makes blood cells. In some cuts of meat (e.g., some steaks) in which the bones have been cut in cross-section, these two types of bone can be easily seen. In those cuts of meat that show the backbone or the ribs, we can see evidence (especially in fish) that vertebrates are segmented animals (see Chapter 2).

Cartilage is a tough connective tissue found throughout the body (Sophia Fox et al., 2009). It covers the ends of bone where they form joints, provides structure to some body parts (e.g., ears and the esophagus), and cushions certain bones (e.g., vertebral discs). Remember that vertebrates have articulated joints, and cartilage helps to make that possible. In humans, damage from overuse of cartilage can result in osteoarthritis (Krishnana and Grodzinsky, 2018). The proportions of the proteins collagen and elastin and protein-carbohydrate molecules, called proteoglycans, give it different physical qualities. As its name implies, elastin provides a degree of elasticity to various organs, such as the skin and the bladder. Collagen is very tough and more difficult to break down than other muscle proteins. Skin and tendons contain a lot of collagen. Muscles that bear a lot of weight or that are used extensively (e.g., legs, chest) have more collagen. The more collagen, the tougher the meat, and cuts of meat with a lot of collagen must be cooked on low moist heat for longer times. That helps to breakdown the collagen. The amount of collagen also increases with age, so animals that are slaughtered earlier in life (e.g., pigs, veal) have less collagen than animals killed later (e.g., cows).

In days long past, meat sections also featured a larger selection of animal organs, including heart, stomach, intestines, kidney, liver, bone marrow, muscle, sweetbreads, and brain. Although the sizes vary from the animals to humans, the anatomy and physiology of these organs are quite similar. Today one might see liver in the frozen meat section or the heart, liver, and crop in a frozen turkey, but most of the other organs are less common today as human food than they once were.

The meat counter reveals a lot about various tissues. Just looking at them, we can see a lot about their anatomy and physiology. Beef or calf liver looks very much like human liver. Both are dark reddish-brown. Livers have four lobes and are highly vascularized. In cross-section, the tissue looks fairly homogeneous because 70%–85% of the liver is occupied by one cell type called parenchymal hepatocytes. Those cells are the workhorses of the liver. They function in metabolism, detoxifying compounds, and protein synthesis (Zhou et al., 2016). They also are part of the innate immune system by secreting opsonins and other proteins that kill bacteria.

Several other organs are still favored by some people. Beef tripe is made from the muscle wall of a cow's stomach chambers. The interior mucosal lining is removed, and the stomach is boiled and bleached so that it turns white. Calves' brains and kidneys are also eaten. Sweetbreads refers to thymus or pancreas. The thymus becomes fibrous with age, and so, pancreas is more commonly consumed. Chitterlings are the small intestines, usually of pigs, that are cooked and eaten. Turkey hearts are found in the giblets package that is placed in the cavity of turkeys that we often enjoy at Thanksgiving. The hearts have four chambers just like all mammals, birds and some reptiles. Beef hearts are also sometimes found in the meat counter.

Birds

Several bird species are included in human diets, including turkeys (*Meleagris gallopavo*), geese (*Anseranas semipalmata*), and pheasants (*Phasianus colchicus*), but chickens are the most popular by far. Humans have kept domesticated chickens (*Gallus gallus*) for meat and eggs for a very long time. The meatiest parts are the flight muscles or breast and the legs, and these are seen as "white" and "dark" meat, respectively. They result from the absence or presence, again respectively, of the protein myoglobin, which is similar to the hemoglobin in our blood. Myoglobin stores oxygen, which is needed for muscles that work continuously. The breast muscles are used only rarely and then for quick bursts of activity and so do not need a constant supply of oxygen. However, the legs and thighs are used for running and walking and need a good steady supply of oxygen. Turkeys are described in more detail toward the end of this chapter.

Seafood

Most supermarkets have an extensive collection of shellfish and finned fish (Seafood Source, nd). Shellfish come from several phyla, including Mollusca and Arthropoda (Vilanova, 2014). Finned fish are all from the phylum Chordata (Tidwell and Allan, 2001).

The major features of mollusks are an immobile exoskeleton (in most) and a mantle that contains their internal organs (Sigwart and Sutton, 2007). Mollusks are segmented animals, but the segments are not easy to see. Common mollusks in our yards include snails and slugs, but the supermarket has clams, scallops, and oysters. Clams are bivalve mollusks: their two shells or valves are held together by a hinge formed by a thick external ligament. The shells are egg-shaped and can be brownish or white or gray. The valves shut when two adductor muscles contract. When the adductors relax, the valves are opened by two ligaments at the top of the clam near the hinge. Inside the shell, a membrane called the mantle surrounds the viscera. The viscera include a heart, gills, and stomach. The foot enables the clam to dig into the sand or mud up to 60 feet deep. Clams are filter feeders and eat planktonic algae. The two siphons protrude from the valves into the water. One siphon is used to suck water and algae down to the clam and through its gills. Food particles are trapped, and the water is expelled through the other siphon.

Squid are another kind of mollusk. However, the body of the squid has evolved considerably from that of the clam. The mantle is the long tube that forms the body. It still encloses the viscera, but it is elongated and more muscular. The foot has also evolved into a set of two tentacles and eight arms. The shell has been reduced to a chitinous "pen" that supports the mantle and anchors various muscles. A siphon is used for very quick movements. Water is sucked into the mantle cavity and expelled in a "jet" from the siphon. Inside the mouth is a chitinous beak that is used to tear food apart. Squid have three hearts. The main heart has three chambers. Two smaller hearts supply the gills. Squid, like their close relatives the octopuses, are very advanced animals. For example, some squid hunt in groups and actively communicate. They have highly developed eyes that are similar to those of vertebrates, but the lens is hard and cannot change shape as human lenses do. Squid can also change the color of their skin to avoid detection by predators.

Arthropods have flexible exoskeletons, and articulated joints can be seen easily on lobsters as they move around a fish tank in many supermarkets. The articulated joints are visible even on the frozen lobster tails. Arthropods are also segmented. In fact, they have combined some segments into larger body sections, such as a cephalothorax (the head and middle section) and the abdomen. Those segments are also very visible in the lobster tails. Crabs, lobsters, and shrimp are crustaceans (Zrzavý and Štys, 1997). For example, crabs have a thick shell called a carapace, short tail, and 10 legs, including two pinchers. The rear-most pair of legs is adapted for swimming. As a general rule, males and females can be distinguished by the shell flap that folds under the crab from the rear. Males have a narrow flap, and females a wider flap. Like humans, crabs are omnivores. They eat most anything. Unlike humans, they have a hard exoskeleton. To grow, they must periodically shed and replace their hard exoskeleton. During that period, they are particularly vulnerable to attack. They are also vulnerable to humans who eat them as "soft-shell crabs." When the body of a cooked crab is opened, the internal organs can be easily seen. The mushy yellowish material is the hepatopancreas, which functions as the liver and pancreas for the crab. The feathery white finger-like projections at the front are the gills. Between the gills is the heart. The blood of crabs is blue because the oxygen-carrying molecule is the copper-based hemocyanin. Just in front of the heart is a part of the stomach called the cardiac chamber. It contains "teeth" that grind up food as a part of the digestive process.

Like humans, fish are in the phylum Chordata. They have a backbone. There are many thousands of species of fish, but the ones typically found in a supermarket come from about 12 species. The characteristics of fish are well known. They are streamlined so they can move effectively in the water with their fins. Their flesh can be white (cod, *Gadus* sp.), pink to red (salmon, *Salmoniformes* sp.), or dark red (tuna, *Thunnus* sp.), depending on the amount of myoglobin in the flesh. Often, the fish are presented as fillets, which are the large muscles from the flanks of the fish. Some, such as swordfish (*Xiphias gladius*) are sold as steaks. In those cross-section slices, the channel for the backbone and spinal cord are quite visible.

Animal Products

We humans enjoy several food products from animals, such as eggs and dairy products. Interestingly, these foods are derived from items that support animals during their early development.

Eggs

Eggs are a familiar item. They are "designed" to provide proteins and energy to a developing chick, but they are a favorite food item for many other animals, including humans. Many animals lay eggs, including fish, turtles, many reptiles, and birds. The eggs we eat are mostly from chickens, though duck and quail eggs are also sometimes available. Humans have been eating eggs for millennia. Chickens are thought to have been domesticated in Southeast Asia and India nearly 7,500 years ago. The average hen lays about six eggs per week. Eggs for sale are rarely fertilized and so cannot ever develop into an adult animal.

The first thing to notice is that they are covered by a fragile shell. Eggs vary somewhat in color from white to brown, but the shell color has no relationship to the nutritional value of the egg inside. The shell is made of calcium carbonate and has many small pores to allow air and moisture to pass through it. A very thin covering called the cuticle protects the inside of the egg from bacteria. Once the shell is cracked, the egg white and yolk can be seen. But if you look more closely, there are additional parts. Just inside the shell are two thin protein membranes. Keratin is a significant component of these two layers. This very tough protein also is used in hair, fingernails, claws, and hooves. An air space forms between the two membranes, usually at the larger end of the egg. The egg white is mostly water and protein, and a lot of that protein is albumen. However, there are actually about 40 other proteins in the egg white. The yolk doesn't just float freely inside the shell and white. It is anchored in position by some twisted proteins "ropes" called the chalazae that can be seen in the egg white. The yolk itself is contained in yet another membrane called the vitelline membrane. The yolk serves as a source of stored energy and nutrients for the developing chick, and so it contains lots of protein, some fats, and several important minerals and vitamins. All of these make eggs a nutritious food for humans too.

Cooking eggs is an excellent demonstration of the denaturation of proteins (Exploratorium, 2023). The egg proteins normally exist in a three-dimensional conformation in an aqueous environment. But certain cooking procedures disrupt those conformations and cause the proteins to stick together. For example, cooking eggs causes the proteins and water floating in the egg to move faster and increases the number of collisions between them. Those collisions damage the proteins and cause it to begin to lose its normal three-dimensional structure. As the cooking progresses, more and more proteins open up and get tangled together. The result is that the liquid egg is turned into a solid rubbery structure. Those same proteins can be damaged by beating. Making a souffle or meringue involves what is called surface denaturation of the proteins. The beating action adds air to the liquid, and again, the three-dimensional proteins lose their structure and form a semisolid mass. If a hard-boiled egg is cooked too long or if there is a lot of iron in the cooking water, the surface of the yolk will turn slightly green. That color results from the reaction of sulfur and iron compounds in the egg with the yolk. The green yolk is safe to eat.

The yolk also contains emulsifiers, which allow oil and water to mix. This ability is key to making hollandaise sauce and mayonnaise. Some emulsifiers in the yolk are proteins that have hydrophobic and hydrophilic amino acids. The hydrophobic amino acids tend to sort into the oil phase, and the hydrophilic ones tend to seek the aqueous phase, and these properties allow the water and oil phases to form an emulsion. Another egg emulsifier is lecithin. This phospholipid has a charged group (phosphate) on one end, and a longer, hydrophobic tail of fatty acids. Just like the protein emulsifiers, the two different groups of lecithin allow the oil and water to form an emulsion.

Milk and Milk Products

Milk is produced by the mammary glands of all female mammals and is used to feed young animals. Thus, this emulsion or colloid in a water base contains valuable lipids, sugars, proteins, vitamins, and inorganic elements to support the growth of infant animals (Willett and Ludwig, 2020). Even human babies begin with breast milk until they can eat solid food. Milk and milk products are widely consumed by humans, especially in more northern regions. In addition, a wide variety of milk products are available, such as yogurt, cheese, butter, and kefir. Modern milk is pasteurized to kill bacteria and fortified with vitamins A and D. Willett and Ludwig (2020) acknowledge the benefits of milk and milk products, especially for young children. However, they are less convinced of the overall value to adults.

Milk and dairy products are important sources of energy, proteins, fats, and nutrients (e.g., calcium and vitamin D). Milk is particularly valuable for young mammals. The main energy source in milk is the sugar lactose. Lactose is a combination of two sugars, glucose and galactose. Thus, it is a disaccharide, and a specific enzyme, lactase, cuts the bond between the two sugar molecules so that lactose can be digested (Forsgård, 2019).

Lactase is found in the intestines, and its levels are high at birth, but diminish over time after weaning. This is also true in humans. However, about 30% of humans retain lactase even into adulthood, but the percentage differs in cultures. Only about 10% of those in Southeast Asia retain lactase, but that percentage is over 90% in Scandinavians. Lactase-persistent individuals inherit a dominant gene mutation that allows lactase to continue to be produced into adulthood. Milk is a rich source of nutrients for humans, and possession of a gene that allows its utilization would be a strong selective factor.

Those without the enzyme are referred to as lactose intolerant. After consuming a lactose-containing product, those individuals suffer multiple gastrointestinal symptoms, such as bloating, diarrhea, and abdominal pain. Fortunately, the food industry has provided a variety of lactose-free products.

Cheese has been eaten by humans since the late Stone Age (Eberle, 2022). Each cheese owes its flavor, color, and texture to the many microorganisms it contains. Even though the cheese may begin its life with a relatively simple starter culture, it may have a complex mixture of microorganisms by the time it is ripe (Pierce et al., 2021). The different microbes in each type of cheese are responsible for the distinctive characteristics of that cheese.

Plant Products

The produce department can be nearly as colorful and attractive as the garden where the fruits and vegetables grew. The sights and smells are wonderful, and the brightly lit, chilled counters are regularly misted to ensure the carrots and Brussels sprouts look their best. With modern transportation, there are now no real seasons for most foods. Fresh strawberries, for example, used to be available only in the spring. Now even in December, they come from greenhouses or faraway places, such as Chile.

Fruits and Vegetables

Surprisingly, the fruits and vegetables in the supermarket represent only a small part of the plant world, at least from a botanical standpoint. No member of the lower divisions can be found at all. In fact, almost every item in our diet comes from a single division: the angiosperms. And of those, we eat only the flowering plants from the subdivision, Anthophyta (Knee, nd). The flowers aren't always very noticeable, but all of the fruits, vegetables, grains, and nuts come from plants with flowers. Humans have been very successful in manipulating plants to improve their characteristics as foods. Wild cabbage (*Brassica oleracea*) is a great example. This native of coastal and southern and western Europe began as a hardy small plant. Today it has several cultivars, including cabbage, broccoli, cauliflower, kale, Brussels sprouts, collard greens, Savoy cabbage, kohlrabi, and gai lan. These are so different in appearance that it is hard to believe that they are the same genus, let alone same species.

Actually, it's not surprising that we would concentrate on this group for food. Those plants and the parts we eat were made by nature to provide energy and nutrients to the plants. Their fruits and tubers have dramatically more stored energy than those of the other groups. In some cases, they evolved to "bribe" animals into disbursing the seeds. For example, birds, bears, and other animals eat berries, but the seeds survive the trip through the animal's digestive system. They are then deposited with a bit of fertilizer at some distance from the original plant.

Roots

Root crops are a significant part of the vegetable section. However, not all "roots" there are really roots. Some are parts of a plant, such as stems, that grow underground but are not technically roots (Herben and Klimešová, 2020). True roots include taproots, which are the main vertical root structure, such as beets (*Beta vulgaris*), carrots (*Daucus carota*), and parsnips (*Pastinaca sativa*). Tuberous roots are enlarged regions of roots that are used for storage and include sweet potato (*Ipomoea batatas*). Some underground structures are actually modified stems. Bulbs, such as onions and garlic, are short stems with fleshy leaves that store food. Rhizomes are main stems that grow underground horizontally. Stolons are similar to rhizomes, but they grow from a stem. Ginger and turmeric are both stem tubers (i.e., storage organs derived from underground stems). Potatoes and yams are tubers that grow on modified stolons and are used by the plant to store starch.

Grains

One of the most important plant products we consume is grain. We eat grain all day in multiple forms. Cereal, toast, or pancakes in the morning. Bread for a sandwich at lunch. Maybe with a beer and a few corn chips on a weekend. Cookies for afternoon snack. Cocktail after work. Rice or rolls for dinner. Cake for dessert. A nightcap. Grains are the seeds of different grasses. Corn, wheat, barley, rice, oats, sugar cane. The seed consists of the three main parts. The bran is an outer shell that protects the seed and includes most of the fiber. The endocarp is essentially carbohydrate to provide energy for the germinating plant. The germ is the growing embryo. Whole grains include all three parts. In processed grains, the bran and germ are milled off to leave only the endocarp. Whole grains are usually considered to be more nutritious than processed grains but not always (Wenner Moyer, 2013).

Grasses have been an important food source for a long time. Some dinosaurs ate grass. Grasses also played an important role in the evolution of mammals. They are the dominant plants of many areas on each of the continents, especially the dry interiors. They spread into many areas as the dense forests receded, and their unique characteristics allowed a number of animals to evolve, including the larger hoofed mammals and smaller burrowing animals. Large fields of grasses made possible the great herds of grazing animals. Importantly, the grasses grown at the base and sometimes even underground and send leaves up. Thus, grazing animals do not kill the growing tip of the plant as happens with other plants. This also lets us mow the lawn without killing the grass. Grasses and some other plants contain cellulose and lignin, two substances that humans cannot digest. Cow and other ruminants have bacteria in their stomach that digest the cellulose in the grass down to its component sugars. The cow does the rest. Interestingly, cows and other grazing animals have developed heavily crowned teeth that can deal with phytoliths, small glass-like particles in the grass. Eating grass is a bit like eating sandpaper (Strömberg et al., 2013).

For centuries, humans have been part of the evolutionary process. They have selected plants with exaggerated fruits or other edible parts. For example, the corn we eat now has changed considerably from the small ears that the Native Americans in Mexico originally cultivated long ago. Maize or corn (*Zea mays mays*) is a member of the grass family *Poaceae* (Kistler et al., 2018). It was initially domesticated from teosinte (*Z. mays parviglumis*), a wild grass, in Southern Mexico about 9,000 years ago (Piperno et al., 2009) and spread throughout the Americas. It reached the United States about 4,000 years ago. The maize was subjected to an enormous amount of selection by humans to transform it from a simple grass with very small grains to a major food crop with the large grains of today (Tian et al., 2009). The grain now lacks glumes, which surround the seeds of other grains, and the seeds remain attached to the cob. These features limit the ability of modern-day corn to grow without human assistance now.

Maize was the subject of a landmark genetic study by Barbara McClintock. She was studying ears with different colors of kernels, including white, purple, and brown (Ravindran, 2012). Until that time, scientists believed that the DNA of genes was linearly arranged on the chromosomes. However, in 1950, McClintock turned that idea on its ear by showing that small bits of DNA move from place to place to affect the expression of genes. The concept of movable genes (jumping genes or transposons) was not immediately accepted, but it eventually earned McClintock the 1983 Nobel Prize in Medicine or Physiology. Today transposons are widely recognized in all species, including humans.

Other grains or grasses are also important. The wheat (and other) flour that we find at the supermarket contains starch and several proteins. Some proteins are water soluble (e.g., albumin, globulin, and proteoses), and others are not (e.g., glutenin and gliadin). When flour is mixed with water, the soluble proteins dissolve. The solid remaining solid mass is primarily the insoluble proteins, which form long strands and give the bread its texture. The strands of glutenin and gliadin together are called the gluten. Different wheat strains have different amounts of these proteins and so different baking characteristics.

Of course, humans in most societies throughout history have discovered another use for the starch in grains: alcoholic beverages. Beer, sake, and many distilled spirits all begin with one or more grains. For example, bourbon must be made from a grain mixture that is at least 51% corn. The rest is wheat, rye, and/or barley. The grain is allowed to germinate to release enzymes that transform the starch to sugar. Yeast cannot metabolize starches, but once the enzymes have changed the starch to sugars, the yeast are ready to go to work. They use ("eat") the sugars as a food source and ferment the mixture to yield alcohol and carbon dioxide. Finally, the product of fermentation is distilled to isolate a fraction of the liquid, containing the alcohol and certain other products.

Seaweed

There is an old cliché that states that the exception makes the rule. Earlier we noted that humans eat almost exclusively flowering plants from the division Anthophyta. Seaweed is one exception. Seaweeds are algae, and evolutionarily, they appeared on Earth long before the flowering plants. Various types of seaweed have been used in foods for some time. These are all different forms of red, brown, or green algae. In many cases, they are simply ingredients and, so, are not noticeable. For example, agar has been used as a thickening agent in ice cream and other dessert applications. Carrageenan comes from a red algae called *Chondrus crispus*. They are widely used in the food industry for gelling, thickening, and stabilizing products. Some seaweed foods are more recognizable. Nori is a dried seaweed from *Pyropia yezoensis* and *P. tenera* that is used as a wrapper for sushi rolls. It has a distinctive flavor. The green algae *Monostroma* and *Enteromorpha* are used as herbs in some meals. Wakame is a brown alga, *Undaria pinnatifida*, that is often used to make salads. Unfortunately, it is also an invasive species (Epstein and Smale, 2017).

Mushrooms and Fungi

Mushrooms are not plants, but they are often found in the fruit and vegetable section. We eat mostly only the small fruiting bodies that push up through the dirt to disperse the spores. The main parts of mushrooms are out of sight underground. In the most common mushrooms (*Agaricus bisporus*)—the ones on pizzas—the spores are produced by the gills on the underside of the cap. The spores can be easily seen by cutting the stem out of a mushroom and leaving the cap, gill side down, on a sheet of paper. The spores will collect on the paper overnight.

There are many types of mushrooms (Li et al., 2021). Many of the mushrooms we eat are from the same species *Agaricus bisporus*. This species includes the small round white ones used on pizzas and the larger brown portabellas and crimini mushrooms, in which the gills on the underside of the cap have yet to open. These mushrooms decompose wood, grass, leaves, and other material. In this way, they recycle plant material for other organisms to reuse. Among the primary decomposers are the oyster mushrooms (*Pleurotus*) and shitake mushrooms (*Lentinula edodes*). Once the material is partially decomposed, other mushrooms join the process, including *A. bisporus*. The morels (*Morchella elata*, *M. esculenta*, or *M. semilibera*) are different. These ascomycetes have a honeycomb appearance due to the network of ridges with pits as their caps. They establish beneficial relationships with the roots of plants they live with: they provide nutrients to the plant, and the plant provides energy to the mushroom. This group also includes chanterelles (*Cantharellus cibarius*), truffles (*Tuber* sp.), and boletus (*Boletus* sp.). Chanterelles belong to the genera *Cantharellus*, *Craterellus*, *Gomphus*, and *Polyozellus*, and are yellow or gold in color and have no true gills.

Several other types of mushrooms might be found. Maitake mushrooms (*Grifola frondose*) grow at the base of trees and can be very large. Shitake mushrooms (*L. edodes*) are originally from East Asia. Enoki mushrooms (*Flammulina velutipes*) grow on the stumps of the Chinese hackberry, ash, mulberry, and persimmon trees. Beech mushrooms (*Hypsizygus tessellatus*) grow on beech trees. King trumpet mushrooms (*Pleurotus eryngii*) are the largest species of oyster mushrooms. Black trumpet mushrooms (*Craterellus cornucopioides*) have no separation between the cap and stem. Hedgehog mushrooms (*Hydnum repandum*) are commonly known as the sweet tooth, wood hedgehog, or hedgehog mushroom, and are a basidiomycete fungus. Porcino mushrooms (*Boletus edulis*) grow in forests and form symbiotic relationships with trees.

Of the many species of fungi, one is used by humans far more than any other. Yeasts are very common in the environment. For thousands of years, various strains of *Saccharomyces cerevisiae* have been used to make bread and alcoholic beverages (beer, wine, hard liquor). The dry packaged yeast in the supermarket can be reactivated by adding them to warm water. A little sugar will really rev them up. After an hour or two, the mix will be frothy as the yeast eat the sugar and release carbon dioxide. The carbon dioxide leavens bread and cakes and creates the alcohol in beer, wine, and spirits (once it's distilled).

Thanksgiving Dinner

Now that we have toured the supermarket, we can take our purchases home for further close-up examination. So let's see what we can learn while preparing for a great American holiday tradition. A typical Thanksgiving dinner comprises turkey, stuffing, mashed potatoes, cranberry sauce, pumpkin pie, bread, and wine or beer. We can look at some of these in more detail.

The obvious "guest of honor" is the turkey. Domesticated turkeys (*Meleagris gallopavo domesticus*) are the same species as wild turkeys. They are thought to have been domesticated about 2,000 years ago in central Mesoamerica, but another domestication event has been identified at 200–500 AD in the southwestern United States. Speller et al. (2010) examined mitochondrial DNA and found that all current turkeys arose in central Mexico in the 16th century.

Turkeys come frozen or fresh and can be prepared by roasting or frying. In any of these cases, the bird arrives without head, feet, or feathers, and the viscera has been removed from the body cavity. The "giblets," which include the heart, crop, liver, and neck, are wrapped and packed in the cavity. The crop or gizzard is a tough muscular organ that grinds seeds and other hard foods. It sometimes includes small stones that the bird swallowed to aid in the grinding. A crop is found in many birds, alligators, and crocodiles. Even some dinosaurs had crops to promote digestion of tough foods.

Birds, such as turkeys, are good models for much of human anatomy (see Krieger, 2010, for an excellent description of the dissection of a chicken with photographs). For our imaginary "dissection," we will begin after the turkey has been cooked, but will reference some things that could have been seen before the cooking.

In preparation for slicing the turkey, the wings and legs are often removed. The joints of the bird are obvious and after cooking, they are quite easy to dislocate from the body at the hip and shoulder joints. Cooking will have degraded some of the connective tissues that held them in place during life. Dislocating those joints also reveals the cartilage. This bright white rubber-like structure cushions the joints between bones (e.g., the knee). It's harder than muscle, but softer than bone. It is made of several materials, including glycosaminoglycans, proteoglycans, collagen, and elastin. In addition to padding the ends of bones where they join, it forms much of the nose, ears, and other structures in the body.

If we look under the skin of the thigh and drumstick (or lower leg), it's easy to see the muscle structure underneath. The large muscle group in the front of the thigh is the quadriceps. Those in the back are the hamstring muscles (i.e., semitendinosus, semimembranosus, and biceps femoris muscles). Those muscles of the chicken correspond to those in the human thigh. The lower leg has more muscles, since it has to move back and forth and also rotate the foot. The three major muscles are the gastrocnemius (the very large muscle mass just below the back of the knee) and the soleus in the back of the leg and the anterior tibialis in the front. It's important to remember that muscles need a structure to act on, and they work only by contracting and relaxing so they work in pairs against each other. (We will examine the actions of muscles, tendons, and nerves in greater detail in Chapter 9.) That structure is provided by the bones. Turkey bones are similar to human bones. The thigh bone is the femur. The drumstick has two bones. The larger one is the tibia. The fibula is quite small in a turkey. Muscles are connected to the bones by tendons. In humans, the Achilles tendon in the lower leg connects gastrocnemius, soleus, and other muscles to the calcaneus or heel bone. Tendons connect muscle to bone, and ligaments connect bone to bone. Separating the thigh and drumstick at the knee also reveals the cartilage at the knee and some ligaments.

Of course, not every aspect of turkey anatomy is the same as in humans. For example, the sternum is the bone in the middle of the chest. It provides an anchorage for the chest muscles (i.e., the pectoralis major, pectoralis minor, serratus anterior, and subclavius) so that we can move our arms inward. In birds, the sternum is greatly enlarged for attaching the major flight muscles. The pectoralis group allows the bird to flap its wings downward. These structures can easily be seen in the chicken or a Thanksgiving turkey. The sternum or breastbone is large and contains a significant cartilaginous component. The two large portions of white meat are those pectoral muscles, and attached to each is also a very strong tendon.

There are many recipes for stuffing, but most contain savory vegetables (e.g., celery, onions), bread crumbs, and spices, with chicken stock and eggs for binding. The vegetables add flavor to the turkey, and the melting fat from the meat drips into the stuffing. Many people also chop up the giblets and add them for additional flavor.

As noted earlier, the starch granules in potatoes are of two types, and both have limited solubility in water. Amylose is linear and dissolves in water at about 131°F. Amylopectin is branched and quite insoluble in water. The granules are coated with protein, and the potato cell walls are made of cellulose, hemicellulose, and pectin. When the potatoes are cooked, the hemicellulose breaks down to allow water into the cells. The granules absorb water, the amylose dissolves in the water, and the starch granules swell and break. The potatoes need to be dried before they are mashed, or they will become watery. Milk and butter are usually added to the mashed potatoes. The milk protein casein acts as an emulsifier and, with the melted butter, slows the leakage of amylose from the broken starch granules to yield a more pleasing consistency.

An explanation of how these elements of a traditional Thanksgiving dinner are digested is included in Chapter 8. Each of the major components is broken down to its simplest parts to be absorbed by the intestines and recycled into new materials for energy or building of new parts.

Supermarket Systematics

A supermarket might not be thought of as a great place to study biology, but in this brief review, we have seen that, in addition to a great selection of foods, a modern supermarket offers an interesting opportunity to explore living things. A number of species (particularly marine species) that are not part of the normal environment for many of us can be seen close up. Those can supplement the species that we have in our homes and backyards. Furthermore, the items in the supermarket allow us to see something of the structure and organization of those organisms, whether plant or animal.

As we have seen, humans eat a lot of different foods, but those foods come from a relatively small number of major animal groups. Now we are at the checkout counter, and we can see one more aspect of biology. The people who run supermarkets understand human animal behavior very well. While we have been hunting and gathering, they have laid a trap to get us with spontaneous purchases by putting small items near the checkout counter (Winkler et al., 2016). The hunter becomes the hunted.

So in our walk around the supermarket, we have seen many of the foods humans eat and learned a bit about them. We eat a lot of different foods, but they come from a relatively small number of major animal groups. Now we are at the checkout counter, and we can see one more aspect of biology. The people who run supermarkets understand human animal behavior very well. While we have been hunting and gathering, they have laid a trap to get us with spontaneous purchases by putting small items near the checkout counter (Winkler et al., 2016). The hunter becomes the hunted.

5

Fellow Travelers

Killing each separate louse is a tedious business when a man has hundreds. The little beasts are hard and the everlasting cracking with one's fingernails very soon becomes wearisome.
—Erich Maria Remarque. *All Quiet on the Western Front*

We tend to think of ourselves as individuals, but we are actually a collection of organisms. Most obvious are temporary pests, such as fleas, scabies, crabs, chiggers, lice, and other small invertebrates, such as those described by Remarque. Some are occasional visitors, but others remain with us. The relationships are mostly harmless or a nuisance. Less obvious are internal parasites, such as tapeworms, hookworms, *Giardia*, *Entamoeba histolytica*, pinworm, and flukes. But there are many, many more.

In fact, we are outnumbered by perhaps 10 to 1 if you compare the number of organisms living on us to the number of our own cells. Bacteria and viruses cover our skin and fill our gut. We floss our teeth to limit the buildup of dental plaque, which is a biofilm produced by oral bacteria. Our gut bacteria are intimately linked to our health. The Human Biome Project has begun to define this relationship, but the buying public is already convinced. It spends a great deal of money on probiotics even though their efficacy is unproved. Babies pick up the flora and fauna that populate their intestinal tract from their mothers during birth. Infants born by cesarean section obtain their gut microorganisms from other sources, such as hospital opportunistic organisms.

Interestingly, others have left their mark on our genes by the genetic remnants of historic parasites, and we have taken advantage of some of those borrowed genes. Up to 8% of the human genome is made up of old viral sequences. We have found some of them to be extremely valuable. For example, the activity-regulated cytoskeletal protein (Arc) is very closely related to a virus sequence. Arc is required for our ability to learn and remember. Retroviruses and long terminal repeat retrotransposons have similar

The Biology of Us. Gary C. Howard, Oxford University Press. © Oxford University Press 2024.
DOI: 10.1093/oso/9780197664797.003.0005

organizations. Retrotransposons only lack a functional envelope gene that retroviruses have. The similarities suggest a common ancestry. Some retroviruses cause cancer. Others are important in human development. For example, retroviruses seem to be deeply involved in placental evolution.

In this chapter, we will look at the living organisms that call us home. We will begin with the temporary visitors on the outside and work our way further inside and end with viruses that invade our genes and stay for a lifetime and more.

Visitors That Drop in for a Bite

There isn't much to like about mosquitoes. At best, they are a nuisance, buzzing around our head and grabbing a blood meal from us while leaving itchy red bumps on our skin (Figure 5.1). At worst, they transmit a number

Figure 5.1. Mosquito. A female *Aedes aegypti* mosquito preparing to bite. Diseases transmitted by this type and other mosquitoes kill millions of people each year.

Reproduced from Gathany, J. (2006). This 2006 photograph depicts a female *Aedes aegypti* mosquito as she was in the process of seeking out a penetrable site on the skin surface. Centers for Disease Control and Prevention. http://www.publicdomainfiles.com/show_file.php?id=1352017 6814273#google_vignette. Public domain license.

of diseases, including malaria, Zika virus, West Nile virus, lymphatic filariasis, eastern equine encephalitis, tularemia, Chikungunya virus, and dengue fever virus. Diseases carried by mosquitoes kill nearly a million people each year and certainly far more than any other animal.

These diseases vary in their severity. For example, most of those who are infected with West Nile virus will not even realize it. About 20% will develop symptoms that include a fever. However, 1% will experience a serious and sometimes fatal neurologic illness. Chikungunya is rarely fatal, but its joint pain can be excruciating and debilitating and may last for weeks. There is no vaccine, and treatment is limited to managing the pain. Mosquitoes also transmit diseases to animals, including dog heartworm, West Nile virus, and Eastern equine encephalitis. Zika is also usually mild. Symptoms, such as fever, rash, conjunctivitis, and joint pain, may last days to months. The incidence is not clear, since many patients may not feel bad enough to seek treatment. Most seriously, Zika infection of pregnant women has been associated with microcephaly and profound neurological defects in newborns.

As with many other challenges, global climate change is extending the range of disease-carrying mosquitoes and spreading the risk of those diseases into new areas (Asad and Carpenter, 2018). Malaria was once endemic in the southern United States, but it was wiped out in the early 1950s with insecticides, draining pools of water and window screens. However, the number of cases has grown in more recent years as more Americans bring it back from areas where it is still endemic. Others, however (e.g., Franklinos et al., 2019), urge caution because many other factors might be involved. Mordecai et al. (2017) examined models of the seasonal and geographic ranges of *Aedes aegypti* and *A. albopictus* and the rates of Zika, Chikungunya, and dengue. Their models show that transmission occurs at 18°C–34°C, and thus, transmission is probably limited to about three months each year in temperate zones. The good news is that that period may be too short to support major epidemics.

Most people would probably have a hard time coming up with any good reason for mosquitoes to exist. From our perspective, they are little more than pesky hypodermic needles with wings. But if we could overlook our knee-jerk response to them and look at them objectively, we might find some grudging admiration if not affection. So what are mosquitoes? They are a large group (nearly 3,600 species) of small flies. They have one pair of wings, a pair of halters, six legs, and elongated mouth parts. Their bodies, legs, and wings are long, slender, and even dainty. Male and female mosquitoes feed on fluids from plants, but females are further adapted for piercing skin. The easily visible feature of the proboscis is the labium that encloses the rest of the mouthparts. Inside are the biting parts that also form a channel for blood to be drawn up into the body. Mosquitoes are attracted to humans by two chemicals that we routinely exhale: carbon dioxide and 1-octen-3-ol or octenol, which is a breakdown product of linoleic acid. They also sense other chemicals in our sweat. They also tend to prefer humans with type O blood, heavy breathers, lots of skin bacteria, high body heat, and pregnant women.

When a mosquito lights, it begins using its labium to look for a likely spot to pierce. The factors in the choice are not clear. When the mosquito bites, it injects a tiny bit of saliva into the wound. The saliva is an amazing combination of fewer than 20 main proteins. First, it introduces an anesthetic so that we do not feel the pain of the bite. Second, several proteins defeat the normal clotting of blood and keep the vessels dilated. Finally, some of those proteins influence the immune response to the mosquito. So the next time you swat at a mosquito at a picnic, remember at least some of the amazing aspects of theses dangerous but amazing animals.

Ticks are another animal that stops by for a quick blood meal and then drops off. They have long been a common problem with dogs and other domesticated animals. More recently, they have become an increasing problem for people. The blood meal that they take is annoying, but it's what they leave us with that is dangerous. Too often, they leave other invaders that stay with us and cause serious diseases. Ticks are arachnids, and so, they are in the same family as spiders. They have eight legs and their cephalothorax and abdomen are fused. They are covered with a hard surface called a scutum, and the mouthparts are formed into a beak. These parasites feed on the blood of various animals, including mammals, birds, and reptiles. Ticks appeared about 100 million years ago, and they were a pest of dinosaurs. Specimens of *Cornupalpatum burmanicum* and *Deinocroton draculi* were found embedded in amber and entangled with a dinosaur feather (Peñalver et al., 2017).

Ticks carry many bacterial, parasitic, and viral pathogens and sometimes even more than one at a time (Madison-Antenucci et al., 2020). Lyme disease is caused by *Borrelia burgdorferi*, and is a serious disease. It's also the most common vector-borne disease in the United States. The cause is the bacterium *B. burgdorferi* or sometimes *B. mayonii*, which is transmitted by the bite of black-legged ticks (deer ticks, *Ixodes scapularis*). It causes fever, headache, fatigue, and a skin rash. It can also spread to joints, heart, and nervous system. The early symptoms usually resolve with treatment, but its chronic manifestations last for years and can be quite debilitating (Feder et al., 2007). Lyme disease is the best known, but there are many others. Rocky Mountain spotted fever is carried by the American dog tick. Tularemia is caused by the bacterium *Francisella tularensis* and is transmitted by American dog ticks and other vectors. Anaplasmosis is caused by a bacterium and transmitted by deer ticks, which also transmit Lyme disease. Babesiosis is a protozoan infection that is often found with Lyme disease. Ehrlichiosis is caused by *Ehrlichia chaffeensis*. Powassan virus disease is caused by a flavivirus. It is similar to relapsing fever. Tick-borne relapsing fever is a bacterial infection that causes recurring bouts of fever, headache, muscle and joint aches, and nausea. Two other diseases are similar. *B. miyamotoi* disease is also carried by ticks, and louse-borne relapsing fever is carried by lice.

More concerning is that the range of tick infestations is increasing. These observations pose serious public health considerations. The diseases carried by ticks cause real damage, and our ability to control tick populations and to detect and treat those infections are limited at this point. With global warming, the ranges of ticks are increasing. This means that they are likely to be a significant concern in the years ahead.

Visitors That Stay for a While

Lice are small, flat, wingless insects and obligate parasites. Like all insects, they have three body segments. The segments of the thorax are fused, but the seven in the abdomen are clearly visible. Of the more than 1,500 species of lice, only two live on humans: head and body lice (*Pediculus humanus capitis* and *P. humanus humanus*, respectively) and pubic lice (*Pthirus publis*).

Head and body lice are closely related, but do not normally interbreed. Head lice live exclusively on the human scalp. They cause itching, are unsightly, and eat blood, but they do not carry disease. They feed several times a day and inject an anticoagulant when they bite. They are typically treated by combing out the nits, which contain the eggs, and appropriate shampoos. Body lice lay their eggs on clothing. Body lice are the "little beasts" noted by Remarque in *All Quiet on the Western Front*. They can transmit disease-causing pathogens, including typhus (*Rickettsia prowazekii*), trench fever (*Bartonella quintana*), and relapsing fever (*B. recurrentis*). They are not common in developed countries, except in areas of poverty with poor hygiene, overcrowding, and a lack of clean clothes. Treatment involves laundering clothes and bed linens in hot water and drying on a hot cycle.

Fortunately, we cannot see some of the organisms that live on our bodies. Eyelash mites or demodex are tiny animals that live at the base of our eyelashes. Mites are in the arachnid group with spiders. Their elongate bodies have two segments with four pairs of legs attached to the first ceph-alothorax segment (Lacey et al., 2011). More than 100 species are known, but only two infest humans. *Demodex folliculorum* lives in a cluster in the hair follicles, and *D. brevis* lives alone in sebaceous glands. They are about one-third of a millimeter in length and transparent. They just eat our dead skin cells and some of the oil that we all secrete. In that sense, they help to keep our skin clean. They occasionally let go of our eyelashes or other hairs and walk around us, especially at night. In almost every case, mites are harmless. Problems result then they get out of control (Lacey et al., 2009). Increased populations of *Demodex* are associated with rosacea and pityriasis folliculorum, a disease caused by too much makeup and too little cleaning of the face. In some cases, the mite is also associated with blepharitis, an inflam-mation of the eyelid.

Internal Parasites

The incidence of internal parasites in humans in the developed world has been greatly reduced in the last century. However, many parasites still infect others around the world and cause a great deal of suffering and death, and some intestinal parasites still infect humans in the United States (Kucik et al., 2004). The infectivity of parasites depends on a number of factors (Leggett et al., 2012).

Pinworms (*Enterobius vermicularis*) are nematodes or roundworms. More than 30% of children worldwide are infected. They live in the cecum of the large intestine, and females emerge each night to lay up to 15,000 eggs on the perineum. The eggs can infect others in the household. The worms typi-cally cause only itching and sleep disturbances.

Giardia (*Giardia lamblia*) is a flagellated protozoan and second only to pinworms for infections in the United States. Giardia is spread by the fecal–oral route, but it is resistant to chlorine in tap water and is often found in rivers and streams. The life cycle comprises two stages. Cysts are passed in the feces. Those are ingested by a new host, and each cyst produces two trophozoites that attach to the wall of the intestines. It causes gastrointestinal issues, including nausea, vomiting, malaise, flatulence, cramping, malabsorption, diarrhea, and weight loss.

Humans are host to two species of hookworms. As the name implies, *A. duodenale* infests Europe, Africa and Asia. *A. americanus* is found in the Americas, but recently has been found in Africa, Asia, and Pacific Islands. Hookworm was endemic in the southern United States until around 1900. Indoor plumbing and the wearing of shoes greatly reduced its incidence. A larval form enters through pore, hair follicles, or other openings and uses the blood system to get to the alveoli. After breaking through the alveoli, they travel up to be swallowed so they can reach the small intestines, where they attach to the wall. Blood loss as the worms feed can caused microcytic hypochromic anemia and other problems.

Entamoeba histolytica is a protozoan that looks like an amoeba. Estimates are that 10% of all humans are infected. Most are asymptomatic, but 100,000 die each year. Its lifecycle is similar to Giardia, but the trophozoites can burrow through the intestinal wall to the bloodstream and on to the liver, lungs, and brain. It can cause bleeding and colitis and malaise, weight loss, abdominal pain, diarrhea, and fever. Those that escape the intestines can cause liver abscesses.

Malaria is caused by a single-celled microorganism. Five species cause malaria, but the most common is *Plasmodium falciparum*. *Anopheles* mosquitoes transfer the organisms to the patients in the saliva of their bite. The parasite is in the saliva of the mosquito. From the bloodstream, they travel to the liver, where they mature and reproduce. While malaria is not endemic to the United States, it is one of the most important diseases in the world. In 2020, there were 241 million cases of malaria worldwide and 627,000 deaths. However, with climate change, there is some concern that the distribution of the *Anopheles* mosquito may increase to put more Americans at risk.

Fungi

Fungi are everywhere, and in most cases, they are harmless. However, even normally harmless fungi can occasionally cause diseases, especially in immunocompromised individuals. Fungal infections can be superficial, subcutaneous, or systemic. Those categories are not exact. Some superficial colonizations can also be systemic (e.g., aspergillosis, pneumocystis pneumonia, candidiasis, mucormycosis, and talaromycosis). Systemic infections often begin in the lungs. Subcutaneous infections often begin with breaks in the skin.

Fungal infections can be very serious, and few drugs and no vaccines are available to treat fungal infections (Brown et al., 2012). Histoplasmosis is caused by a soil fungus *Histoplasma capsulatum.* It typically infects the lungs and can be fatal if untreated. Cryptococcosis is caused by another soil fungus *Cryptococcus neoformans.* It infects the lungs and the brain and is also potentially fatal. Immunocompromised individuals, such as AIDS patients, are particularly vulnerable.

Other fungal infections are less serious. Tinea refers to several fungal infections. Athlete's foot is known as tinea pedis. Several genera, including *Trichophyton, Epidermophyton,* and *Microsporum,* cause the condition. Symptoms are typically itching, scaling, and cracking between the toes. It can also affect other areas of the foot or hands. Infections usually result from exposure to the organism on wet floors around swimming pools and in locker rooms. Jock itch is tinea cruris, and it occurs more often in males than females. Its symptoms are red patches and itching in the groin area that do not usually involve the scrotum. *Tinea capitis* is scalp ringworm. It usually affects small children and is highly contagious. It features a red, scaly, itchy rash on the scalp, loss of hair, and rashes on other body parts. *T. unguium* is an infection of the finger or toe nails. The nails become thickened and deformed. *T. corporis* is body ringworm. In this condition, ring-shaped rashes appear anywhere on the body or face.

Bacteria

Most of our experiences with bacterial infections are unpleasant. Most sore throats are caused by viruses and will clear up by themselves, but strep throat is a more serious infection caused by group A *Streptococcus* bacteria. Untreated it can result in damage to the kidneys and rheumatic fever and damage to the heart. The acne that many teenagers suffer is a complex process, but bacteria are a key participant. The main bacterial component is the anaerobic bacterium *Propionibacterium acnes.* Infants often experience the pain of ear infections. They can also be caused by either viruses or bacteria. *Streptococcus pneumoniae* or *Haemophilus influenzae* are common culprits. In the infection, the Eustachian tubes are blocked, which results in a painful buildup of fluid in the middle ear.

Bacteria are much more than a few painful memories from our childhoods. Describing bacteria requires a lot of adjectives. Small. Single-cell. First. Ubiquitous. Extraordinary. Harmless. Deadly. They are very small and mostly single-cell organisms. They can be found just about everywhere on Earth, and they thrive under extreme conditions of temperature and pressure where it's hard to believe that anything can survive. They were likely among the first life on Earth. Most are harmless. Some are beneficial and even necessary partners with us, but a few are pathogens and even deadly. They come in many shapes, including spheres, rods, and even spirals. Bacteria are prokaryotes. They generally lack the internal membrane structures that characterize eukaryotes, such as a nucleus and mitochondria. Their genome is found in chromosomes and in small lengths of DNA called plasmids.

Good dental care helps to protect against many oral bacterial infections. The incidence of dental cavities has been greatly reduced by early dental treatments, such as fluoride in the water and tooth sealants. Many bacterial species are involved in tooth decay. Found that each sample contained 70–400 species, and they differed between individuals and even within the same individual (Simón-Soro et al., 2014). *Lactobacilli* were found mostly in dentin cavities, and *Streptococci* were found in 40% of decay in enamel. Dental plaque is a sticky deposit that forms on teeth (Marsh, 2006). If it isn't removed, it turns into brown or yellow tartar. It covers the teeth, but builds up between teeth and along the gumline. It is, in fact, a biofilm. Biofilms are mainly bacteria, but can also contain fungi. About 70% of the dry weight of the plaque is bacteria. The rest is polysaccharides and glycoproteins. About 1,000 bacterial species are associated with plaques, and they use the plaque to protect themselves from viruses or bacteriophage (Winans et al., 2022). Altogether, they form a hard deposit that is a major cause of tooth decay and gum disease. Regular hygiene, such as brushing teeth and flossing, help to contain the biofilm, but optimal control includes a visit to a dentist or dental hygienist to get the teeth cleaned once or twice a year. Poor oral health may also affect other diseases, including atherosclerosis, diabetes mellitus, Alzheimer's disease, rheumatoid arthritis, and more (Scannapieco and Cantos, 2016). The inflammation may be the key to all of these.

While most bacteria are neutral or beneficial to humans, some cause really serious diseases. These include tuberculosis (*Mycobacterium tuberculosis*), anthrax (*Bacillus anthracis*), tetanus (*Clostridium tetani*), pneumonia

(*S. pneumonia, Staphylococcus aureus, H. influenza*, and more), cholera (*Vibrio cholerae*), botulism (*C. botulinum*), meningitis (*S. pneumoniae*, group B *Streptococcus, Neisseria meningitidis, H. influenzae, Listeria monocytogenes*), gonorrhea (*Neisseria gonorrhea*), and bubonic plague (*Yersinia pestis*).

Microbiome

The human microbiome comprises bacteria, archaea, fungi, viruses, and other microbes that live on and in us humans (Ursell et al., 2012). The term "microbiome" was coined by Lederberg and McCray (2001). The microbiome was described by Turnbaugh et al. (2007) as having two parts. The core microbiome comprises the set of genes of organisms in and on a human. These are fairly constant from one human to another. The variable microbiome includes the genes of those organisms that vary from habitat to habitat and according to the health or sickness of the host, lifestyle, environment, and transient microorganisms. While most of the microorganisms in our body are found in the gut, they are not the only ones. Other parts of our body also have collections. The skin, mouth, nose, ears, vagina, and urinary tract have their own microbiomes.

The microorganisms on the skin depend on the physical characteristics (e.g., moist, dry, sebaceous) of the skin (Byrd et al., 2018). The bends of the elbows and the feet are moist, and the dominant species are *Staphylococcus* and *Corynebacterium*. The most prevalent bacteria on sebaceous sites are *Propionibacterium* spp. Fungi are less common everywhere on the body. One group of fungi (*Malassezia* spp) was found on the body core and arms. Feet show Ma. *Malassezia* spp., *Aspergillus* spp., *Cryptococcus* spp., *Rhodotorula* spp, *Epicoccum* spp, and more. One fungus, Candida (*Candida albicans*), lives in dark, warm, moist places, such as the folds of skin and the genital area. It also grows on the tongue, throat, or inside cheeks. Also called thrush or candidiasis, it's not common, but usually affects those who are immuno-compromised from AIDS or chemotherapy.

The main cache of microorganisms is in the human gut. The gut typically has about 1,000 bacterial species with perhaps 2,000 genes per species. That means that the gut contains about 2 million genes, or 100 times the human genome (Gilbert et al., 2018). It totals about 5 pounds (2.3 kg) of material. All of those cells secrete signaling proteins and other molecules that allow them to interact with other cells (ours and others) and they are integral to our immune system. Altogether, the microbiome has probably 300 times more genes than we humans have in total.

We begin with none of these, or so it is thought. More on that later. We emerge from the birth canal into a sea of microorganisms that populate our mother, and her microbiome becomes ours. The microbiome is important, even critical, to our health, and we need it from the start. The sugars in mother's milk must be broken down by bacteria in the baby's gut before they can be absorbed and used for energy by the baby. Babies born by caesarean section do not receive the bacteria from the mother, and some scientists believe that those children are more vulnerable to asthma and type 1 diabetes than children who experienced a vaginal birth.

Interestingly, the organisms that make up the microbiome change with time. Our microbiome changes over the first few years. We gain additional bacteria from our mother's milk and the environment. By about age three, it is relatively stable, but it can still be affected by changes in our environment, diet, stress levels, and drugs (e.g., antibiotics). The genetic diversity of the microbiome is under active investigation (Barud and Pollard, 2020). The forces that determine variability in other populations are also are work here (e.g., mutation, recombination, drift, and selection) and those changes can have profound effects on the effectiveness of drugs.

The microbiome has been linked to multiple disorders, including autism, anxiety, obesity, how we respond to drugs, the effectiveness of chemotherapy, and even how we sleep. For example, the microorganisms might encourage obesity by affecting our appetite, the gases produced by digestion, and how well we use our food, or by influencing our immune system or inflammatory processes. Changes in the microbiome are translated throughout the immune, endocrine, and nervous systems and correlate with many diseases (e.g., inflammatory bowel disease, cancer, and depression). Just as the microbiome may affect our health and well-being, so it might be tweaked to help us overcome various disorders. In fact, a multi-million-dollar industry has developed around probiotics in the hope of doing just that. Probiotics are foods or supplements that contain live microorganisms. Probiotics are heavily advertised to help with gastrointestinal ailments. Eating them is meant to maintain or restore intestinal health. Most, such as yogurt, are healthy foods and certainly not harmful. The gut's acids and enzymes are meant to protect us from bacteria and other microorganisms that are eaten. Few survive this harsh environment. Yet, some do survive and live to colonize the gut although it is far fewer than promised. In addition, bacteria contribute a number of natural products to our system, and their metabolites may be beneficial.

One treatment involving modifications to the microbiome has proved highly effective. Infection with *Clostridioides difficile* results in severe diarrhea and inflammation of the colon. About 500,000 cases are reported each year. They are particularly dangerous for older people: one in 11 people over 65 who get it will die within a month. Antibiotics are often used to treat this disorder, but the infection often returns within a couple of weeks. A more effective treatment involves a fecal transplant, in which fecal material from a healthy donor is implanted in the colon of a patient (Shogbesan et al., 2018). The donated material colonizes the patient's intestines and overgrows the *C. difficile* to prevent it from regaining a foothold.

So are we born with a microbiome or do we gain it at some point? For many years, the fetus was assumed to grow in a sterile environment. Bacterial infections are a in utero (Brokaw et al., 2021). Chorioamnionitis refers to an infection of the chorion and amnion, the membranes that surround the fetus and hold the amniotic fluid. In most cases, this requires the fetus to be delivered as soon as possible. Recent studies have thrown this concept of a sterile fetal environment into some doubt. Several studies reported evidence for bacteria in the gut of the fetus. Bacterial DNA has been detected in meconium and amniotic fluid samples. Those discoveries suggest that the uterus is not sterile. Stinson et al. (2019) examined 50 women who were scheduled for a nonemergency cesarean section and who had no evidence of uterine infection. They found that all meconium (the first stool from a neonate) and most amniotic fluid samples contained bacterial DNA. *Propionibacterium acnes* and *Staphylococcus* spp. were the most common. Bacteria have been found in the meconium (Chu et al., 2017; Durack et al., 2018). Others have showed that bacterial colonization is limited to the intestines in utero (Rackaityte et al., 2020). Mishra et al. (2021) looked for evidence of microbes in utero and found low levels of 16S rRNAs that indicated bacteria. They also found bacterial strains (e.g., *Staphylococcus* spp., *Lactobacillus* spp.) in fetal tissues that might be involved in fetal immune priming. They also discovered bacteria-like structures in the fetal gut lumen at 14 weeks.

However, the existence of microbes in utero is still controversial. Samples are hard to obtain and are usually quite limited in quantity. There are many opportunities for contamination, especially in the delivery room. Kennedy et al. (2021) argued that no study to date has sufficiently controlled for these multiple sources of noise and potential contamination, and thus, any report of fetal microbiota should be carefully scrutinized.

Viruses

There are a lot of viruses. Some estimates are that 10 nonillion (10 followed by 30 zeros) viruses are on the Earth. Viruses are obligate parasites. To replicate themselves, they must get inside a living cell so they can hijack that cell's machinery and obtain energy. With their small genomes and high replication rates, viruses deal with lots of mutations, some of which make them more infective. And they have had billions of years to evolve new ways to succeed.

The plant viruses are astoundingly economical. For example, tobacco mosaic virus contains one RNA molecule and 2,130 copies of a simple protein. These components automatically assemble themselves to produce a finished virus particle. Being so simple, the virus is very stable. It can live on the floor of deserted tobacco barns for 100 years and still be infective. Infection involves mechanical damage to the leaves or stems of tobacco plants. Even breaking the fragile hairs on the underside of the leaves is sufficient. Once in the cell, the virus readily uncoats itself so that the RNA can begin producing copies of itself and the coat protein using the host systems.

One of the most fascinating is the bacteriophage T4 (Figure 5.2) that infects the bacterium *E. coli*. The phage looks a bit like a lunar lander spacecraft (Maghsoodi et al., 2019). The DNA genome is packed into the head at the "top" of the structure. A long tubular sheath connects the head to the baseplate. Attached to the base plate are six articulated long fibers of "legs." The long fibers bind to receptors on the surface of the *E. coli* and draw the baseplate close to the bacterium. Short fibers in the bottom of the baseplate then connect to the cell surface, and the sheath contracts to drive long tail piece or "needle" through the cell membrane. The viral DNA then migrates from the head and into the cell to begin the infection.

Viruses infect humans by several mechanisms. Some hitch a ride with those pests that bites us for a blood meal, such as mosquitoes, mites, ticks, and fleas (mentioned above). However, others must find a susceptible cell to attack. Proteins on the surface of the virus interact with a receptor protein on the surface of the host cell. The specificity of this interaction defines what organisms or cell types that the virus can infect. Nonenveloped viruses (those without a lipid outer shell) have mechanisms to penetrate the cell

Figure 5.2. T4 bacteriophage. This virus attacks *E. coli* bacteria. The legs contact the bacterium and draw the main body down to the cell. Once bound, the long tube contracts to inject the viral genome (housed in the top) into the bacterium.

Reproduced with permission from the U.S. National Science Foundation. (2004). The bacteriophage T4 is preparing to infect its host cell. Purdue University and Seyet LLC. https://www.nsf.gov/news/mmg/mmg_disp.jsp?med_id=51293&from=mn.

The animation is based on both recent discoveries and extensive earlier work by a large number of investigators. A full list of contributors is available at the conclusion of the animation.

membrane to allow the virus access to the cytoplasm (Dimitrov, 2004). Enveloped viruses have a lipid outer shell with proteins protruding through the lipid layer (Harrison, 2008). Once the virus binds its receptor, the viral lipid shell fuses to the lipid membrane of the host cell. Membranes are made of lipids and proteins. The lipids are fluid and can fuse with each other. Viruses clip along a membrane until they bind a protein that they recognize that is embedded in the lipid layers. The viral lipid coat and the lipid membrane fuse with the help of fusion proteins (F Cohen, 2016). This action allows the virus to enter the cell by endocytosis, and after entry, the virus releases its genetic material to begin the infection. For example, the SARS-CoV-2 spike glycoprotein (Figure 5.3) binds to the angiotensin-converting enzyme 2 (ACE2) receptor on the cell surface (Ni et al., 2020). Lungs express large amounts of ACE2 receptor and are exposed to the virus-filled air that

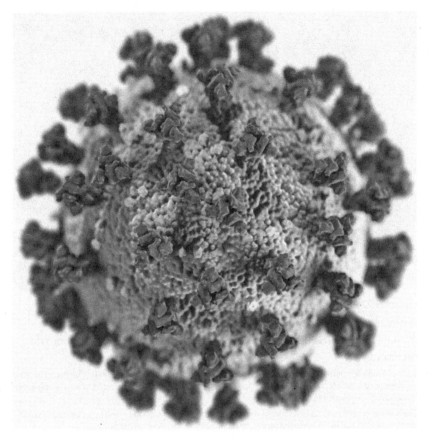

Figure 5.3. Coronavirus. A novel coronavirus, SARS-CoA-19, caused the pandemic that began in 2019. The structures on the outside of the virus are the spike proteins that are involved in binding to human cells.

Reproduced from Eckert, A., and Higgins, D. (2020). 23312. Centers for Disease Control and Prevention. https://phil.cdc.gov/details.aspx?pid=23312. Public domain license.

we inhale. Other cells that express ACE2 (e.g., myocardial cells, proximal tubule cells of the kidney, bladder urothelial cells, and enterocytes of the small intestine) are also vulnerable to SARS-CoV-2 attack.

Viruses That Come and Go

Today vaccinations are mostly taken for granted. Pediatricians begin a series of vaccinations early on in infants, and compliance is very good (Table 5.1).

Table 5.1 Vaccination Rates in the United States by Age 24 Months

Vaccination	Rate
DPT (diphtheria, pertussis, tetanus)	80.7%
MMR (measles, mumps, rubella)	90.8%
Chickenpox	90.2%
Polio	92.6%
Hep B (hepatitis B)	90.6%

Source: NCHS (2015)

Many people get annual vaccinations for influenza, and the recent coronavirus pandemic might eventually require a similar annual vaccination. Even with all of these successes, some viruses continue to haunt us. The best example is the common cold, which results in millions of cases each year. Several viruses cause colds. The most common are the rhinoviruses, but there are others, including respiratory syncytial virus, human parainfluenza viruses, adenovirus, common human coronaviruses, and human metapneumovirus. Fortunately, most of these resolve themselves within 7 to 10 days, but a lot of misery can be packed into those days.

Americans have not always so enthusiastic about vaccinations. In the 18th century, smallpox caused deadly periodic epidemics. The only known way to avoid the disease was to stay away from others. Some people advocated for a new procedure that had been used in China since the 10th century. Variolation involved inoculating a person with material from an active smallpox pustule (Belongia and Naleway, 2003). In many cases, immunity resulted, and the patient survived. Unfortunately, in some cases, the inoculated person became sick and died. In 1721, an epidemic broke out in Massachusetts. Some argued that variolation was against God. The modern smallpox vaccination began with Edward Jenner, who noticed that milk maids rarely came down with smallpox. They often contracted cowpox, a more benign but related virus that resulted in immunity from smallpox. Since that time, smallpox has been eradicated from the Earth, and there is no longer a need for smallpox vaccination.

A more recent emergency occurred in the 1950s. Polio infections reached about 60,000 per year. Children were especially hit hard. They suffered from paralysis and death, and the iron lung became a tragic symbol of the era. Jonas Salk developed a vaccine with killed virus, and the virus was approved rapidly. Polio has also become a huge success story. Cases resulting from the native virus are confined to Pakistan and Afghanistan. In fact, there are far more cases of polio resulting from mutation of the live attenuated virus to a pathogenic state than there are from native virus.

Viruses never sit still. They continue to evolve, driven by mutations that change their properties. The coronavirus that caused COVID-19 is just one more in a long and continuing line of emerging viral threats. As humans continue to affect the climate, ecosystems, and biodiversity, new and re-emerging viruses are likely to appear.

Existing viruses cause many outbreaks. Influenza remains a serious threat. The 1918 strain caused 20 million deaths. That strain is assumed to have been an avian influenza. The noroviruses periodically cause intestinal distress and are very contagious. Dengue virus, Japanese encephalitis virus, West Nile virus, Chikungunya virus, yellow fever, and Zika virus frequently cause outbreaks with significant morbidity and mortality. Hantaviruses are RNA viruses that live in rodents and are transmitted through urine and feces. They cause hemorrhagic fever with renal syndrome and hantavirus pulmonary syndrome. Ebola and Marburg cause hemorrhagic fever and are extremely contagious and lethal. Oropouche orthobunyavirus and Mayaro virus are candidates for epidemics from South America.

Viruses That Stay

Most viruses infect our cells, complete a few rounds of replication, and move on to another host as our immune system begins to identify and kill them. However, some viruses come to stay. These include human immunodeficiency virus (HIV), the herpesviruses, the hepatitis viruses, and oncoviruses.

Since the 1980s, the HIV pandemic has killed millions of people worldwide. Once the viral and host membranes merge, the virus is released into the host cytoplasm, and its protein coat dissolves to release the RNA genome. Like all viruses, HIV cannot reproduce on its own. It hijacks the host cell machinery to make copies of its RNA and the viral proteins. The single-stranded RNA is copied into a double-stranded DNA by a viral enzyme called reverse transcriptase. This step is characteristic of a group of viruses called retroviruses. The viral DNA can then access the nucleus and integrate itself into the host DNA. From there, the viral DNA is transcribed into viral RNA that can be translated into viral proteins. Thus, the viral RNAs and proteins are ready to be assembled into complete new virus particles that can be budded off from the cell to infect other cells. The infected host cells can only produce virus for about 2 days, but they can each yield an average of 250 new viral particles.

HIV also kills cells that at first did not seem to be infected. These so-called bystander killings were unexplained for some time until Doitsh et al. (2014) showed that the cells were infected, but the infection failed. However, the failed infection initiated a series of events that resulted in pyroptosis, which killed the cell and released a storm of inflammatory cytokines that attracted more immune cells to the killing site.

Fortunately, powerful drug combinations have transformed AIDS from a lethal infection into a chronic disease. The drugs kill virus that is actively replicating and reduce the levels of virus in the blood to undetectable levels. However, if the drug regimen is interrupted, the virus reactivates, and viral counts can rebound in short order. Many drugs have been developed to attack the viral replication process at each of these steps. Among the first were inhibitors of the reverse transcriptase. Others include integrase inhibitors that prevent the integration of the viral DNA into the host genome. Protease inhibitors interfere with the assembly process. While those drugs are very effective against an active viral infection, they cannot kill the viral DNA that is incorporated into the host genome. For that reason, this latent virus is the great challenge now (Siliciano and Greene, 2011; Larragoite and Spivak, 2019). How can that be killed? Various strategies are being tested to reactivate the virus and kill it (Rodari et al., 2021). The primary strategy has been "shock and kill," in which the virus is shocked into activity and drugs are used to kill the virus. The trick has been to find an effective drug that will cause reactivation of all of the latent virus. This remains the most serious aspect of completely curing HIV infection.

HIV is only one of a number of viruses that have a latent phase in humans (Traylen et al., 2011). Among the other viruses are human papillomavirus (warts) and herpesviruses, such as herpes simplex 1 and 2, Epstein-Barr virus (infectious mononucleosis), hepatitis B (hepatitis) and C (possible non-Hodgkin's lymphoma), and herpes zoster (chickenpox and shingles).

The herpesviruses are a large family of more than 100 viruses, including several that commonly infect humans (e.g., herpes simples 1 and 2, varicella zoster, cytomegalovirus, and Epstein-Barr) (Whitley, 1996). All of the herpesviruses form latent infections that last indefinitely. These viruses have a similar structure. The DNA genome is covered by a protein capsid with an icosapentahedral shape. Next is another layer of protein called the tegument, and surrounding everything is an envelope made of a lipid bilayer studded with glycoproteins. The herpesviruses depend on latency and reactivation to establish a persistent infection without killing its host's cells (J Cohen, 2020). To accomplish this, they have evolved several intriguing strategies. The viral DNA is not integrated into the host genome. Instead, the virus forms circular copies of its DNA that are called episomes. When the cells divide, the episomes are connected to chromosomes by a virus-encoded tethering protein. This mechanism ensures that the episomes are distributed to each daughter cell. The virus also produces long noncoding viral RNAs that are transcribed in the reverse direction as normal (called antisense) to limit the production of viral proteins and enable the virus to avoid detection by the immune system. Finally, the viral genes are also "turned off" by modifications to key chromosomal proteins called histones. The histones are reversibly chemically modified by the addition of methyl and acetyl groups to help maintain latency.

The herpesviruses cause several diseases. Herpes simplex 1 (mostly oral) and 2 (mostly genital) differ considerably in their DNA sequence, but they cause similar symptoms (Whitley and Roizman, 2001). Transmission requires intimate contact. They replicate in epidermal cells and result in a characteristic vesicle ("cold sore") on base of reddened skin caused by blood-engorged capillaries. The infecting virus travels along the sensory nerves to the dorsal root ganglia, where it establishes a latent infection. Reactivation results in new lesions at the skin and/or mucosal surfaces.

A primary infection by herpes zoster or varicella zoster causes chickenpox (Weinberg, 2007). The virus is highly contagious, and approximately 95% of adults have been infected. Those numbers are now reducing due to an effective vaccine. Latency is established in the dorsal root ganglia. The virus can be reactivated to travel down the sensory nerves, resulting in shingles, which manifests as a vesicular rash with acute neuritis. Fortunately, an effective vaccine against shingles is now available.

The γ-herpesvirus Epstein-Barr virus (EBV or human herpesvirus 4) is a common viral infection in humans. It was originally found during examination of cells from a patient with Burkitt's lymphoma by electron microscopy (Young and Rickinson, 2004). This childhood tumor is quite common in sub-Saharan Africa. At some point, most people become infected. The virus is spread through bodily fluids, such as saliva. Infectious mononucleosis is caused by EBV and also other viruses. The symptoms include extreme fatigue, fever, sore throat, head and body aches, and swollen lymph nodes. The symptoms usually resolve in a couple of weeks, but they can remain for months. EBV infects B lymphocytes by a typically mechanism: one of the envelope proteins, g350, binds to a receptor protein, CD21, on the cell surface. A small percentage of the EBV infects the memory B cells and can assume a latent existence that allows it to lie dormant for decades. The virus has been implicated in numerous cancer types, in addition to Burkitt's lymphoma, and immunocompromised patients can suffer relapses of the virus that can be serious and even fatal.

Human cytomegalovirus (CMV) is a member of the herpesvirus family and is also known as herpesvirus-5. It is common in people of all ages (Cannon et al., 2010; De Groof et al., 2021). Nearly one-third of children and more than half of adults by age 40 are infected. However, our immune system keeps the virus in check in most cases so that most people are unaware that they are infected. Like all herpesviruses, CMV establishes a latency or dormant state that persists for the lifetime of the individual (Goodrum, 2016). Those who become immunocompromised or immunosuppressed lose that protection, and the virus can reactivate. For example, the virus can cause transplant rejection and death in transplant patients. Later in life, CMV may cause mucoepidermoid carcinoma and prostate cancer. CMV infection in neonates is a very serious and potentially fatal disease that affects about one out of every 200 babies. About one in five babies with congenital CMV infection will have long-term health problems, involving deafness, learning disabilities, and intellectual challenges.

Some viruses are associated with cancer (Moore and Chang, 2010). In those cases, infections (e.g., human papillomavirus [HPV]) result in the viral DNA being inserted into normal genes that cause them to become cancerous. Others (e.g., HIV, hepatitis C virus [HCV]) simply disrupt the immune system so that the surveillance for cancer cells is less effective. Infection by one of these viruses does not mean that you will get cancer. For example, there are many strains of HPV, but only a very few cause cancer. Epstein-Barr virus is associated with nasopharyngeal and stomach cancers and some lymphomas. Hepatitis B and C viruses are associated with liver cancer. HIV is associated with Kaposi sarcoma, cervical cancer, non-Hodgkin lymphoma, and anal and lung cancers. Human herpesvirus 8 is also found in nearly all cases of Kaposi sarcoma. Like HIV, human T-lymphotropic virus-1 is a retrovirus. It is associated with lymphocytic leukemia and adult T-cell leukemia/lymphoma.

Viral Fossils

Viruses sometimes insert themselves into our DNA as part of their replication strategy. HIV is a great example of this strategy. The virus incorporates itself into the infected person's genome and begins producing its RNA and proteins by hijacking the host's transcriptional and translational machinery. For a small number of cases, the virus enters a dormant state. Periodically, some virus reactivates, but some copies remain dormant for the life of the person.

Retroviruses are those that reverse transcribe or make a DNA copy of their RNA genome so that it can be integrated into our human genome. Once inserted, that DNA sequence is copied back into RNA so that viral proteins can be made to produce new virus particles. In most cases, the viral DNA is inserted into a somatic cell. However, a few are inserted into a germ cell, and so, the viral DNA sequence can be inherited along with our other genes for generations into the future.

Retroviruses have been infecting humans and our ancestors for millions of years. As a result, the human genome includes a significant number of DNA sequences that were once free-living RNA viruses. The exact number is difficult to calculate. Some sequences in our genome are easily recognized as of retroviral origin. These are usually called human endogenous retroviruses (HERVs). However, the word "human" is only loosely correct. They are part of the human genome, but they were incorporated into the genomes of our distant ancestors long before any hominid appeared. HERVs make up about 8% of our total genome (Mustelin and Ukadike, 2020). That might not seem like much at first. To put it into some context, all of the genes that encode for the proteins that form us only account for about 1% of the total genome. So 8% is an astounding large number to come from viruses. More amazing is that the number is certainly much larger. The 8% are clearly viral. However, retroviruses were infecting our ancestors for many millions of years, and random mutations, deletions, recombinations, and other changes have occurred over the millennia so that the viral sequences are less recognizable. Viral insertions from more than 100 million years ago are probably difficult to identify. These genetic "fossils" are called "paleoviruses." Additional viruses can be inferred by their effects on the evolutionary pressures on host genes, and they continue to exert their influence on our own resistance to new viruses (Emerman and Malik, 2010).

It's hard to estimate the number of paleoviruses in the human genome. Many duplications, transpositions, and other events have affected the number. However, one estimate of the number is about 100,000 copies from 31 families (Emerman and Malik, 2010). There are likely many more because most infect somatic cells and are lost at each generation. Parrish and Tomonaga (2016) believe that even more of the human genome had retroviral origins. They place the fraction as one-quarter. They also note that endogenous retroviruses are involved in multiple functions, including placenta formation, resistance to related viruses, and modulation of the innate immune system and pluripotency.

All organisms seem to have retroviral sequences in their genome. Thus, it is likely that the insertion of these genes began very early in the history of living organisms on Earth. Since these viruses have been transferring DNA sequences from one organism to another for all of that time, it's hard to imagine that this mechanism has not been deeply involved in evolution.

Each retroviral insertion adds about 9,500 basepairs of DNA to the genome. Among those sequences are the genes for the gag, pol, and env proteins and additional sequences. These components have been used as raw material to evolve proteins with different functions (Grandi and Tramontano, 2018). For example, RNase H and integrase from the pol gene evolved into parts of our immune system, and retroviruses may have contributed the ability to splice our messenger RNAs. In addition, each insertion has one long terminal repeat at each end that contains various regulatory elements, such as promoters and enhancers, that control the transcription of genes.

Other sequences within the human genome are related to retroviruses. The retrotransposons are an even larger group. They include the short interspersed nuclear elements (SINE), such as the Alu elements. About one million Alu elements are found in our genome, and many of them are involved in regulating transcription and mRNA processing or splicing. They retain some of the characteristics of retroviruses. They periodically transcribe or make a copy of themselves and then paste that copy back into the genome at a different location through reverse transcription. They also can amplify their numbers rapidly, and in fact, they make up over 40% of the human genome. Thus, some scientists have said that humans are more retroviral than human.

Bornaviruses are RNA viruses that have infected animals for tens of millions of years (Gifford, 2021). Their remnants can be found as endogenous bornavirus-like elements scattered throughout many genomes.

By studying these viral fossils, one can gain a better understanding of the epidemics that plagued our evolutionary ancestors. Kawasaki et al. (2021) used the sequences of the three bonaviral genera to determine a comprehensive history of ancient viral infections over nearly 100 million years. Their work chronicles the virus–host coevolutionary history over this extensive period.

One retroviral related sequence is called the activity-regulated cytoskeleton-associated protein (Arc). The Arc protein has homology with the retroviral Gag protein that is required for viral infectivity (Shepherd, 2018). Arc is conserved through many organisms and has multiple functions. Interestingly, mice lacking Arc can learn, but they cannot remember. Thus, Arc seems to be involved in our ability to transform what we learn into long-term memory. Arc is associated with autism and other neurodevelopmental disorders (Parrish and Tomonaga, 2018). How memories are stored in our brain is not really understood. Neurons form connections to each other at synapses, and the number and strength of the connections at the synapses are somehow involved with memories. However, an individual neuron might have connections with a few or more than 100,000 other neurons. As a result, the "code" for memories is not clear at this point. What is clear is that memory storage involves several steps that yield changes to synaptic strength. Signals are quickly sent from the synapse to the nucleus, and specific genes are rapidly transcribed. Those mRNAs are packaged and moved to the synaptic site. Arc was long considered to be a cytoplasmic protein, but Korb and Finkbeiner (2011) showed that Arc also crosses the nuclear membrane. More recently, two groups showed that Arc is involved in a unique transfer of information between neurons. Pastuzyn et al. (2018) expressed the Arc gene in bacterial cells and found that the proteins formed a structure similar to a viral capsid. Ashley et al. (2018) studied extracellular vesicles from the motor neurons of fruit flies (*Drosophila*). The vesicles contained high levels of Arc mRNA. The greater the level of activity, the more Arc mRNA was included in the vesicles. Both groups showed that the vesicles delivered the Arc mRNA to other neurons. Furthermore, mouse neurons lacking the Arc gene took up the vesicles and used the mRNA to make Arc protein. The transfer of vesicles was unidirectional: from motor neuron to muscle. The transfer of the Arc mRNA was unexpected and represents a new mechanism for the transfer of information and material from cell to cell in a manner that viruses have used for millennia. It might have potential for further exploring or treating neurodegenerative diseases.

Endosymbiosis

Perhaps the most amazing "infection" occurred very early on in the history of life, and it was one of the most profound changes in evolution. Primitive life 1.2 billion years ago consisted of single cells, and they were prokaryotic-like. Prokaryotes have no internal membranes, but eukaryotes do. At some point, proto-eukaryotic cells took up other cells that were similar to bacteria or cyanobacteria. They established an endosymbiotic relationship, and result was extraordinary. Those that took up the cyanobacteria began to photosynthesize. They all ultimately took up bacteria that evolved into mitochondria.

This concept was first suggested by the German botanist Andreas Schimper in 1883 during his studies of chloroplasts. Chloroplasts are organelles in plant cells in which photosynthesis takes place. Schimper noticed that the chloroplasts divide and their division looks a lot like that of free-living cyanobacteria. He speculated that the chloroplasts had once been free-living and had entered into a symbiotic relationship with plant cells. In 1905, the Russian botanist Konstantin Mereschkowski first suggested the endosymbiotic theory. In the 1920s, the American biologist Ivan Wallin added mitochondria to the theory. Mitochondria are the organelles that produce energy in all cells. The theory was met with great skepticism, but it remained in the background.

That all changed when a seminal paper by Lynn Margulis (then Lynn Sagan) put the concept on firm experimental ground (Sagan, 1967). She was not a molecular biologist, but she noted that mitochondria and bacteria are about the same size. Furthermore, they are self-replicating and have two membranes and the inner membrane of mitochondria is similar to that of bacteria. The mitochondrial DNA molecule is circular, just like bacterial DNA. Even then, her paper was rejected by more than a dozen journals before being published. The theory received compelling support from Schwartz and Dayhoff (1978), who compared the DNA sequences of mitochondria and bacteria.

Since the original "bacterial infection," mitochondria have evolved into a cell organelle (Boguszewska et al., 2020). They gained transport proteins, a structure that involves cristae, and biochemical pathways for glycolysis and lipid synthesis. Many, but not all, of its original genes migrated to the host genome. So the production of many of the proteins needed by the two organelles is carried out by the normal host system and transported back to them.

In the End

Our "fellow travelers" outnumber of own cells manyfold. Most are benign, and many are helpful and even critical to our own health. We live in relative harmony with them. As long as we are alive, our immune system, skin, and mucous membranes keep those tenants at bay. At times, the balance between them and us shifts so that they can get out of control and cause disease. In addition, as we have discovered in the past decades, many of tenants are critical to our well-being.

6

Inside and Outside

The concept of inside and outside is both simple and profound. It is one of the most basic engineering problems that life had to solve. In fact, life depends on having an inside and an outside. All living organisms struggle to keep desirable things inside and to exclude undesirable things or to corral valuable reactions and products.

Biology offers many examples of the importance of inside and outside, and those examples occur at every level from the cells to tissues and organs to individuals and groups of animals. For example, musk oxen (*Ovibos moschatus*) live in the far North. Their chief characteristics include an extremely shaggy coat and large horns on both males and females. Like many animals, the adults are very protective of their young. When the herd is threatened by wolves or other predators, they form a circle with their horns facing outward (Figure 6.1). This solid front keeps the young calves safe inside. The ring of adult musk oxen demonstrates the importance of the inside and outside in biology, but it is easy to think up other examples.

The problem of inside and outside started with the very beginning of life on Earth. Many scientists believe that life began in water. How it happened is not understood, but it is believed to have begun in small pools of some sort. In a letter to Joseph Hooker in 1871, Charles Darwin wrote about "some warm little pond" that might have contained a mixture of chemicals and was exposed to a source of energy. He speculated that a protein substance might have been created that could change over time. In a classic experiment in 1953, Stanley Miller and Harold Urey showed that basic organic chemicals associated with life could be produced from the exposure of an early Earth atmosphere to electrical discharges that simulated lightning (Miller and Urey, 1959). Many other experiments have since shown that various useful biological molecules result from these mixtures.

While Miller and Urey showed that chemicals important to life can be formed abiotically, those products could easily be lost. The chemicals and chemical reactions that began life might have occurred in small tidal pools. As the pool dried up, the concentrations of the chemicals might have reached

The Biology of Us. Gary C. Howard, Oxford University Press. © Oxford University Press 2024.
DOI: 10.1093/oso/9780197664797.003.0006

Figure 6.1. Musk oxen protective circle. Young musk oxen are protected by the defensive perimeter of the adults.

Reproduced from Adams, L. (2023). Defense strategy. National Park Service. https://www.nps.gov/cakr/learn/nature/mo-defense.htm.

levels sufficient for reactions that could have started life. Heat from the sun or from volcanic rocks might have provided the energy to accelerate the reactions. However, those ponds could be overwhelmed by a rainstorm, flood, or high tide. Or they could have dried up entirely. Other hypotheses have been suggested. For example, the black smoker thermal vents on the ocean floor could have provided energy and a mineral-rich environment. Whatever the mechanism, the nascent life needed a means of protecting its biochemical reactions. And that brought up a need for ways to seal off precious reactants and products so they could be part of further reactions. How they accomplished that remains one of the key questions about the origin of life.

But we need not look back billions of years for the importance of inside and outside. To start with, we can examine ourselves. We have an inside and an outside. Actually, we have several insides, depending on what aspect of us we are looking at. Our skin is part of our outside. It protects us from infections and holds fluids inside us. Our mouth seems like it should be inside, but it is actually outside. Food enters our mouth, but it doesn't really go inside us. It remains outside all the way through until it is eliminated as waste at the other end. Thus, the lumens of our stomach and intestines are really on our outside. Humans are sort of an extended donut. Mathematicians would call us a torus. The real "inside" of us includes our tissues, organs, and muscles, but it excludes the inside of our digestive system.

A defined inside and outside is established early in development. The fertilized egg divides many times to form a small ball of cells. Those cells migrate to the outside of the ball to leave an empty space in the middle. This structure is called the blastula. Soon the blastula invaginates. Imagine a deflated basketball that is pushed in on one side until it almost closes in on itself. The "inside" of the basketball is now surrounded by two layers of rubber. At this stage, the gastrula is surrounded by two layers of cells. The positions of cells in the developing embryo gives information to each of the cells about what kind of cell they will ultimately become.

In all of these cases, the inside and outside are important. Each of our cells has an inside and an outside. Even inside the cell, there are areas that are sequestered by cell organelles, such as the nucleus where our genes are stored and mitochondria where energy is produced. How do living organisms separate inside from outside? Even more than three billion years after Darwin's ponds, controlling products, reactants, enzymes, and optimal reaction conditions a still critical to cells. Furthermore, cells are the building blocks that build larger compartments in which activities need to be sequestered. Cells are the basic building blocks of our tissues and organs. They form part of the structure, and they secrete other structural components to contain activities. They form our skin to separate us from the environment. Cells are components of our blood vessels. They line our stomachs and intestines. They form the blood-brain barrier (BBB). And much more.

How do cells separate the inside from the outside? The various types of separation and how they form are very complicated, at the most fundamental level, they are based on a very simple concept, oil and water, or salad dressing. The dressing is shaken vigorously before adding it to the salad to briefly disburse the oil in the aqueous vinegar. Once the container is allowed to settle, the two liquids separate with the oil on top. Oils "like" to stay with oils, and water "likes" to stay with water. Oils are hydrophobic, and water is hydrophilic. There are ways to get them to stay together. One way is to add an emulsifier. In making mayonnaise or Hollandaise sauce, an oil or butter and water are mixed, and an egg is added as a common emulsifier. Proteins in the egg yolk called lecithins attach themselves to the oil droplets and allow them to remain mixed in the water.

Biological membranes use the same basic idea to create an inside and outside. Before we talk about that, let's review the physical characteristics of each component. The three basic ingredients in this recipe are water, oil (lipids), and proteins.

Water

> If there is magic on this planet, it is contained in water.
>
> –Loren Eiseley (1953)

Water. We bathe in it, make coffee with it, and brush our teeth in it. Many people carry water bottles everywhere. It refreshes us on a hot day. We sail on it and swim in it. We water the lawn. We dodge raindrops and shovel snow. We are mostly water ourselves: about 70%. We can go without food for weeks, but we can survive only a few days without water.

In fact, life as we know it could not have occurred on earth without the almost magical properties of water. Its properties have affected nearly every aspect of life. Several of water's important characteristics can be seen in an ordinary glass of iced tea (Figure 6.2). Iced tea is an infusion of tea leaves by liquid water. Ice floating in the tea cools it, and water droplets condense on the side of the glass. All of these features demonstrate the physical characteristics of water, and our life depends on them.

So what is it about water that makes it so important? Its chemical formula is as familiar as it is simple. H_2O. Two atoms of hydrogen bound to one atom of oxygen. However, it is not the atoms themselves that make water special. It's the way they are joined. The two hydrogens are not symmetrically distributed around the oxygen. The properties of how electrons circle the atomic nucleus caused them to be positioned relatively close together—much like Mickey Mouse ears—rather than on opposite sides of the oxygen (Figure 6.3). This off-center arrangement yields an angle of 108° and makes life possible.

In its atomic nucleus, oxygen has eight positively charged protons, but each hydrogen has only one proton. The negative electrons spinning around the hydrogens are pulled just a bit toward the oxygen protons. As a result, the oxygen side of the water molecule has a slight negative charge, and the hydrogen side has a slight positive charge. The differences in the distributions of the electrical charges are small but more than enough to

Figure 6.2. Ice water. The photograph represents the three phases of water: liquid water, ice, and the drops on the outside result from condensation of water vapor.

Photograph reproduced with permission from Rebecca Howard Valdivia.

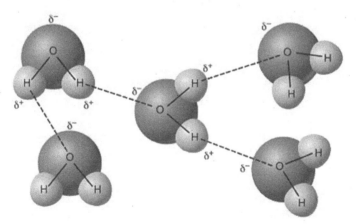

Figure 6.3. The structure of water. The two hydrogens are positioned to one side of the molecule. As a result, the electrons of the two hydrogens are pulled slightly toward the oxygen molecule. As a result, the oxygen has a slight negative charge, and the hydrogen end is slightly negatively charged. Each water molecule has polarity that allows it to weakly bind to other molecules. This interaction explains why water beads up.

Reproduced from the Water Science School. The strong polar bond between water molecules creates water cohesion. U.S. Geological Survey. https://www.usgs.gov/media/images/strong-polar-bond-between-water-molecules-creates-water-cohesion. Public domain license.

give water polarity, and the water molecules attract each other like tiny magnets. That polarity also allows other molecules that have a charge to dissolve easily in water. For example, table salt is sodium chloride (NaCl). The sodium part has a positive charge, and the chloride has a negative charge. In water, the two separate readily and interact with the positive and negative charges on the water molecules. The salt can be called a hydrophilic molecule. It "likes" the water.

The polarity manifests itself in many ways, and some are very easy to see. On a hot day, water droplets condense on the side of a glass to form beads. They are roundish because that tiny bit of attraction between the water molecules holds them together. This "surface tension" also be seen in the bulge of water above a filled glass. The same force causes water to bead up on a car hood and allows a water strider to walk on the surface of water without sinking in.

In addition to polarity, water has several other remarkable chemical and physical properties that are critical for life on Earth. For example, water is the only compound to exist in all three physical forms at temperatures humans can tolerate. Again, we can see this in a glass of ice water. The glass contains both liquid water and solid water (ice). On the outside of the glass, the beads of water that form are evidence of the water vapor all around us that has condensed onto the glass. Once again, the ice in the glass floats. Pure water freezes at 0°C, but it is most dense at 4°C. The density of frozen water is less than the density of liquid water. That allows ice to cover and insulate rivers and lakes from the cold of winter, which allows fish and other aquatic organisms to survive. Water also has a high heat capacity. It takes more heat to raise the temperature of water than of many other compounds. And water holds fewer oxygen molecules than the air. These properties are the reason that is so important to life on earth. Without water, there would be no life as we know it. Oxygen is less concentrated and diffuses more slowly in water than in the air. This property has been important in the evolution of respiratory systems. It also affects nutrient availability and biological diversity in temperate lakes. It also raises the question of how plants and animals, whose cells are essentially bathed in water, can survive subzero temperatures.

Thus, water is an amazing substance. Its physical characteristics are critical for life on many levels. Certainly, one of those is the mechanism for establishing the boundaries around critical biochemical reactions, as we will see below.

Oils and Lipids

The next ingredient is the oil. While water is a single simple compound, there are many kinds of oils and fats, and they have different chemical structures and different properties. Collectively, they are called lipids. Oils are liquid at room temperature, and fats are solids. This difference reflects their different chemical compositions.

Fats have a bad reputation these days. We live in a world awash in calories and short on physical exercise. Obesity, diabetes, and heart disease are epidemic and lead to many deaths. Fats get a lot of bad press. However, fats are just as critical to life as water. They are a very efficient way to store energy: they contain nearly twice the calories of carbohydrates. They provide the raw material for a number of hormones and signaling molecules. Finally, they are a key component of membranes, which we will explore more fully below.

Benjamin Franklin was intrigued by oils. Amazingly, in the late 18th century, he conducted an experiment that provided an estimate of the size of an individual oil molecule.

> If a drop of oil is put on a polished marble table, or on a looking glass that lies horizontally; the drop remains in place, spreading very little. But when put on water it spreads instantly many feet round, becoming so thin as to produce prismatic colors, for a considerable space, and beyond them so much thinner as to be invisible, except in its effect of smoothing the waves at a much greater distance. (Benjamin Franklin 1774)

As Franklin noted, oil spreads out on water, and different colors appear on the surface of the oil sheen. The colors are iridescent and result from the light reflecting both from the surface of the oil and the surface of the water. We commonly see this when gasoline or oil is spilled into a mud puddle. The spreading and colors are due to the physical characteristics of the oil and water. Those same characteristics are critical for life and the ability of organisms to compartmentalize biochemical reactions. It's a simple and elegant as oil and vinegar for salads. More on that later.

Like many of his contemporaries, Franklin was fascinated by science. His experiments with magnetism and electricity made him well known in the world of science, and his natural curiosity and imagination led him to try an interesting experiment with oils. He wanted to measure the size of an oil molecule, and he devised a very clever experiment to do just that. He started with an assumption, which is actually a very good one. He assumed that oil poured onto water would spread out so that its thickness was only one oil molecule thick: a monolayer. With that assumption, he reasoned that he could estimate the size of a single oil molecule.

In the experiment, Franklin poured one teaspoon of triolein, a type of triglyceride that is a main ingredient of olive oil, onto a large body of water. He found that the oil spread over an area of about half an acre. The oil molecules spread out on the water in what he assumed was a single layer. He knew the starting volume of oil, and his measurements gave him two dimensions of the oil on the water. Thus, it was a simple algebraic calculation to find the third dimension, which was the thickness of the oil layer and one molecule. He estimated that the oil molecule was 2.5 nm, a very good estimate from such a simple experiment. In fact, triolein is approximately 2 nm in length.

Triglycerides are among the most important energy storage molecules (Figure 6.4). These simple lipids are made by joining a glycerol, a three-carbon molecule with an -OH group on each carbon, to three long-chain fatty acids. Fatty acids are made of a string of carbons bound together in a long chain. At one end is an acid group (-COOH). An enzyme joins the glycerol -OH to the acid -COOH in a condensation reaction that releases a water molecule (H_2O). When all three of the -OH groups are bound, the molecule is a triglyceride, which has no charge and is quite hydrophobic (it does not like water).

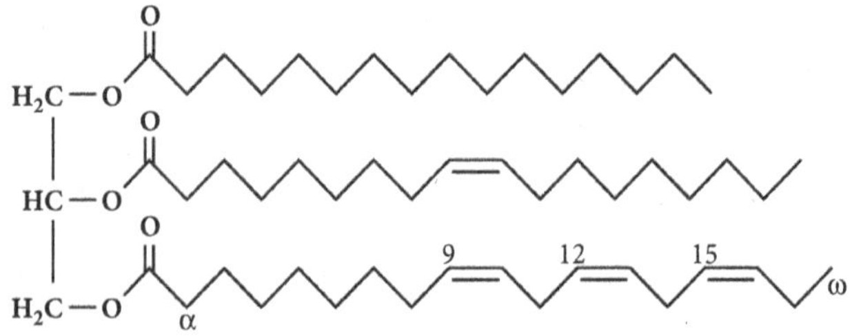

Figure 6.4. Triglyceride. Triglyceride is a major energy-storage molecule. Three molecules of fatty acids are joined to a glycerol molecule. In some lipids, one of the fatty acid molecules can be substituted for another molecule with an electrical charge.

Reproduced from Wolfgang Schaefer. (2005). Fat triglyceride shorthand formula. https://commons.wikimedia.org/wiki/File:Fat_triglyceride_shorthand_formula.PNG. Public domain license.

Not all lipids are as simple as triglycerides. Triglycerides can be modified so that one of the –OHs is not connected to a fatty acid. For example, a phosphate group ($-PO_4$) has a negative charge. Once that chemical group is added, the glycerol has two opposing characteristics. The two fatty acids are hydrophobic and seek other like molecules. The PO_4 group is hydrophilic and seeks other like molecules. Because of their structure, these phospholipid molecules will orient themselves so that the phosphate group is toward water and the fatty acid tails are all together. The result might be for them to form an interface by separating or by forming small spheres. In biology, two layers of lipids will orient themselves so that the lipids are "inside" and the phosphate groups are facing the water. A lipid bilayer membrane separates the aqueous outside of a cell from the aqueous inside of the cell, and it is quite effective in doing this.

Proteins

So far, the discussion has been about water and oil or lipids and how they form a membrane that separates inside from outside. However, the idea of inside and outside can be extended to other biological molecules and particularly to proteins. Proteins are biological polymers—long chains of monomers linked together (Scitable, 2014). In the case of proteins, the monomers are amino acids. The order of the amino acids that are linked into the polymer is important and dictated by the genes that encode the information to make those proteins (Stollar and Smith, 2020). The order is specific for each different kind of protein. That will be made more clear below.

Biology uses approximately 20 different amino acids to make the protein polymers needed by living organisms. As the name implies, each has an amino ($-NH_3$) and an acid ($-COOH$) group. Those are the groups that are linked to form the polymers. But the part of each amino acid that is most important is referred to as an R group, and each R group has different chemical properties that define the overall characteristics of that amino acid. Some are neutral. Some are electrically charged as either positive or negative. Still others are hydrophobic.

Because of the nature of the amino acids, proteins also have an "inside" and "outside." An "average" protein is a long polymer of 300–500 amino acids. That sequence of the 20 different amino acids is called its primary conformation, and in that sense, the protein is just a really long, very thin molecule. But it doesn't stay that way. As soon as it is made and in fact while it is being made, it begins to fold back onto itself (Figure 6.5). It does that in a

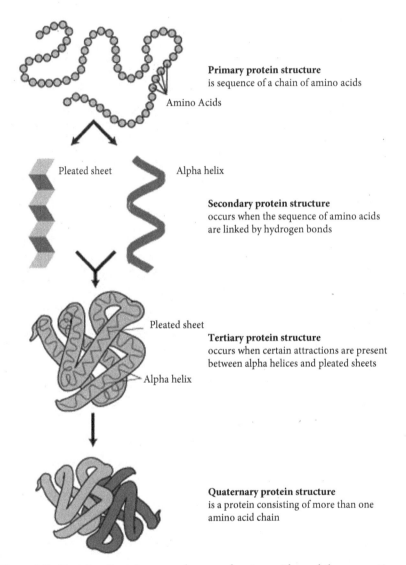

Primary protein structure
is sequence of a chain of amino acids

Amino Acids

Pleated sheet   Alpha helix

Secondary protein structure
occurs when the sequence of amino acids
are linked by hydrogen bonds

Pleated sheet

Tertiary protein structure
occurs when certain attractions are present
between alpha helices and pleated sheets

Alpha helix

Quaternary protein structure
is a protein consisting of more than one
amino acid chain

Figure 6.5. Proteins. Proteins are polymers of amino acids, and the properties of those amino acids cause the protein to assume specific structures. The hydrophobic amino acids tend to be found inside the protein. The three-dimensional structure is responsible for the functions of the proteins.

way that is controlled by the R groups. The R groups that have charges either attract or repel each other. Those that have hydrophobic groups try to get closer together. The result of all of these preferences is that the protein forms a three-dimensional structure. The first set of folds can fold back onto themselves, and even other proteins can be involved in the overall complex. Thus, the structure of proteins is dictated by the characteristics of the amino acids and results in the protein having a hydrophobic "inside" and a hydrophilic "outside."

That three-dimensional structure is extremely important in biology. The function of the protein is tightly linked to its structure. If the structure changes—sometimes only a tiny bit—the function can be changed or lost. In some cases, even the change of a single amino acid can change the structure and the function. That is the case in sickle cell anemia. The protein that carries oxygen in the blood is called hemoglobin. It consists of two alpha (141 amino acids each) and two beta chains (146 amino acids each) (Finch et al., 1973). A negatively charged glutamic acid in the beta chain is replaced by a hydrophobic valine, and this small change causes a dramatic change in the hemoglobin molecule, the shape of the red blood cell, and the oxygen-carrying ability of the hemoglobin (Coleman and Inusa, 2007).

Antibodies also consist of multiple chains of proteins that form a three-dimensional structure (Chiu et al., 2019). Antibodies of the family of immunoglobulin gamma consist of two light chains and two heavy chains that form a protein with two binding sites that recognizes the structure of an antigen.

As with most other things in biology, there are exceptions. While many proteins have a stable three-dimensional conformation, a significant number of eukaryotic gene sequences code for regions of proteins that lack a well-defined structure (Dyson and Wright, 2005). The amino acid sequences of these proteins are enriched in small and hydrophilic amino acids and proline. The functions of these intrinsically unstructured regions are just beginning to be understood. Some form linker regions that connect different domains of a single protein. Others have important functions. The ability to easily assume multiple conformations may be useful in signaling or control processes where processes need to be turned on and off efficiently. By unfolding, they might expose more surface area to ensure additional specificity. Strikingly, unstructured regions are often related to diseases. For example, unstructured regions have been implicated in amyloid deposition, which may be involved with Alzheimer's disease and type II diabetes (Knowles et al., 2014).

Membrane Structure

We now have all of the main ingredients (e.g., water, lipids, proteins) to make a membrane. Lipids and water form a boundary at their interface, but if there is water on both sides, the lipids will form into a double layer with their hydrophilic sides near the water and their hydrophobic, fatty acid tails huddled together in the middle (Figure 6.6). In this way, the membrane can seal off a compartment where specific biochemical reactions can take place. These compartments form cells and the organelles within the cells (e.g., nucleus, mitochondria, lysosomes), and as noted above, this ability to sequester various activities is critical to overall cell function and viability (Honigmann and Pralle, 2016). The membranes are studded with proteins that are anchored in the membrane or even reach completely through the membrane to the other side (Figure 6.6) (Lee, 2001).

Membranes are composed of more than 1000 lipid and protein molecule types that can be modeled as a two-dimensional fluid (Kalappurakkal et al., 2020). Their structure is even more complicated than this large number of components implies. A living bilayer membrane is an asymmetric structure, including lipid synthesis and transfer, membrane traffic and turnover. The membrane also interacts with the cytoskeleton and the extracellular matrix. The proteins embedded in the membrane also complicate the interactions.

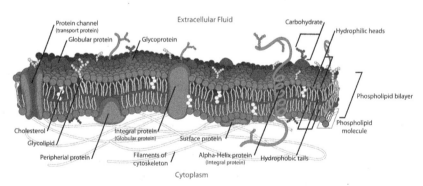

Figure 6.6. Lipid bilayer membrane. Lipids form layers with the charged groups facing the outside (the aqueous phase) and the hydrophobic groups inside. Protein channels enable the passage of various molecules.

Reproduced from LadyofHats. (2007). Cell membrane detailed diagram. https://en.m.wikipedia.org/wiki/File:Cell_membrane_detailed_diagram_en.svg. Public domain license.

Crossing Membranes

Before I built a wall I'd ask to know
What I was walling in or walling out
　　　　　　　—Robert Frost, "Mending Wall" (1914)

A secure barrier is important for any living cell, and it certainly was an early engineering problem for life. The valuable large molecules, made by random reactions, had to be preserved. They could not be allowed to be washed away in the next tide or rainstorm. Those primitive membranes likely involved shorter-chain lipids. Lipid chains of 14–18 carbons offer a good compromise between retaining large molecules and allowing smaller metabolites to pass through.

The lipid bilayer forms a remarkably effective barrier that defines inside and outside for the cells. A few small molecules can cross the membrane, but larger molecules or ones that have a strong electrical charge are often blocked by the membrane. This ability to sequester reactions and products is invaluable for the cells and the total organism. Lipid membranes evolved over time to form a protective envelope around cells, and in more advanced organisms they provide an excellent way to compartmentalize reactions within the cell itself. For example, the lysosome is responsible for breaking down worn-out proteins and other molecules and recycling the parts (Ballabio, 2016). It has a rather low acidic pH and contains a number of enzymes that break proteins apart. The low pH environment is part of the process, and the proteolytic enzymes are effective and need to be contained where they can act on damaged proteins and not other proteins in the cell. A good internal membrane system allows this to happen. In like manner, the DNA that makes up our genes must be protected so that the valuable information is not degraded. The nuclear membrane ensures that this takes place.

But solving one problem can raise another. There are times when the cell needs to move materials or information across the membrane. How can it do that? Cells need to absorb nutrients and to excrete waste materials. They also need to receive and respond to messages from their environment. How can they do these critical functions? Once an impermeable barrier is built, how does anything pass? How do needed reactants cross the primitive membrane? And how are waste products eliminated? How are materials moved from one cell organelle to another as they are processed? How are signals received? For some small molecules (e.g., water, salts), transport through the pores is by simple diffusion. For other larger molecules or substances that need more regulation, active transport mechanisms are used, and some require energy expenditure.

This problem has been solved with proteins. Cells use protein structures that cross the membrane. The proteins are embedded into the membrane by way of their sequence of amino acids and their three-dimensional structure. For example, some proteins can form channels that cross the membrane from the inside to the outside. The domain of the protein embedded in the membrane is more nonpolar than the domains that extend into the more polar environments of the cytoplasm and extracellular fluid. Those parts of the channel inside the membrane are enriched in those amino acids that are hydrophobic. The channel can then allow specific molecules to enter or leave the cell. Other proteins cross the membrane too. They receive signals (e.g., hormones, smells, light) from various kinds of stimuli. The most common are called G-protein-coupled receptors (Rosenbaum et al., 2009). They are the target of a very large number of drugs. Sriram and Insel (2018) cited 134 drugs that target these receptors. Once stimulated, they change their structure so that it also changes on the other side of the membrane. That might release a signal molecule inside that can travel to the nucleus and turn genes on or off to make an appropriate response to the stimulus.

The use of hydrophobic and hydrophilic forces—the oil and water—define the inside and outside of the cell and of different compartments inside. They provide a strong barrier to prevent the loss of needed resources and to protect against outside invaders. Yet the same barrier allows nutrients access to the cell and waste material and secreted products to be moved out. It also allows signals to be received at the cell surface and to be transmitted across the membrane.

To infect cells, viruses must cross membranes, and they use different strategies to do that. Some viruses infect bacteria by directly injecting their genetic material into the cell. But viruses that infect humans, such as HIV and coronaviruses, have lipid membranes that fuse with the cellular membrane to gain access to the cytoplasm. For example, the coronavirus that causes COVID-19 uses its spike protein to attach to the ACE receptor on the surface of cell (Muñoz-Fontela et al., 2020; Scudellari, 2021). Once that connection is made, the virus can position itself so that its lipid outer layer can fuse with the cell. A similar but reverse process occurs when mature virus particles are released from the cell.

Other Lipids

Lipid droplets are the major storage structure for lipids in cells. They are formed by enzymes (diacylglycerol acyltransferase) on the endoplasmic reticulum. The exact process is not known, but several aspects are clear. Walther et al. (2017) suggested that the triglycerides are synthesized by enzymes until sufficient numbers are collected to form a droplet.

Technically, because of the presence of lipid droplets, cells are emulsions, and emulsions are only partially stable systems (metastable). The droplets will remain in suspension for a period, but eventually, they will coalesce into larger droplets and then a continuous lipid phase. The system seeks to minimize the surface tension between the two phases. Small droplets have a larger energy cost, and so, they will coalesce with others to minimize the lipid surface area that is exposed to the aqueous phase.

For lipids to be used, they must be moved from one place to another through a blood system that is aqueous and relatively polar. To accomplish this, lipids are collected into particles that are bound by specific lipid transport proteins. The protein components of these lipoprotein particles help to direct the lipids to receptors in the cell membranes where the lipids can be taken up and processed by the cells (Ference et al., 2020). The lipoproteins are called apolipoprotein (apo) A, B, C, D, E, and H. ApoB is the low-density lipoprotein (LDL) or "bad cholesterol" that doctors warn us about. ApoE is part of the high-density lipoprotein or "good" cholesterol that protects against heart disease. Brown and Goldstein (1976) explained cholesterol metabolism by discovering the receptor for LDL particles. Their work laid the groundwork for the introduction of statins to control cholesterol levels in the blood and won them the 1985 Nobel Prize for Medicine.

Milk is an important food for most humans, at least as infants. It provides proteins and essential fatty acids to the newborn. Fat globules (0.1–15-μm diameter) contain phospholipids, glycoproteins, enzymes, and cholesterol. Whole milk is mostly water, but it also contains some percentage of fat. The fat forms an emulsion (Singh and Gallier, 2017). Milk is an emulsion of fat globules in an aqueous solution. The globules are stabilized by a milk fat globule membrane of phospholipids, glycoproteins, enzymes, and cholesterol. These molecules act as surfactants in the milk. However, in raw milk, the fat will over time separate from the water phase and collect at the top. However, almost all milk is "homogenized." That process breaks the fat into microscopically small particles that remain dispersed throughout the water phase. Chemists would call this emulsified mixture a colloid.

Soaps and detergents are similar. Soaps are made from natural material, and detergents are man-made. Either one, when combined with water, is very effective in destroying bacteria and many viruses (Jabr, 2020). The soap molecules have a hydrophilic head and a hydrophobic tail. The tail associates with the lipids in the cell membrane of bacteria or the lipid outer layer of many viruses and causes them to dissociate from the organism. This disruption of the outer surface is lethal to the microorganism. The action of the soap is very similar to that of a surfactant.

Inside and Outside of Us

Many of the processes within our own bodies need to be segregated away from other activities (e.g., digestion, circulation, pregnancy). In like manner at a molecular level, some activities need to be contained within our cells.

Membranes are critical for cells (Watson, 2015). Membranes surround the entire cell, and they contain most of the cellular organelles, including the nucleus, lysosome, and mitochondria. In addition, a system of membranes called the endoplasmic reticulum and Golgi apparatus is found throughout the cell. This system is critical for maintaining the cellular conditions (e.g., pH, salt concentrations, reactants, enzymes). Those biochemical reactions could not occur without containing the biochemicals at the needed concentrations and conditions. The lysozyme is an excellent example. It is involved in the degradation and recycling of cellular proteins and other components. The lysosome is very acidic and it contains enzymes that dismantle proteins. The acids and enzymes would be toxic to the cell if they were loose in the cytoplasm. However, they work efficiently within the lysosome.

The brain uses large amounts of energy and so needs a continual supply of oxygen and glucose. The blood vessels that serve the central nervous system (CNS) have a unique structure called the BBB) (Daneman and Prat, 2015). The BBB carefully regulates the ions, nutrients, and gases that reach the CNS, and it also protects the CNS from toxins, pathogens, inflammation, injury, and disease. The blood vessels are made up of two types of cells. Endothelial cells form the vessel, and mural cells that sit on the outer surface of the vessel in a matrix of laminins and collagen called the basement membrane. The vessels are joined by tight junctions and lack any of the openings seen in vessels in other tissues.

Ruptures of the BBB result from a stroke, multiple sclerosis, traumatic brain injury, Alzheimer's disease, epilepsy, or other causes. They cause serious damage to the neural tissue. At least part of the cause of the damage is the leakage of the blood-clotting protein fibrinogen and its related form fibrin (Petersen et al., 2018). While these proteins are critical for clotting of blood, they can also initiate an inflammatory process that can further damage neural tissue near the BBB rupture.

The stomach has a major role in digestion of our food (Stumpp et al., 2015). To do that, it secretes acid and has enzymes (e.g., pepsin) to process the food into molecules that can be absorbed later in the intestines. This topic will be explored in great detail in Chapter 8. However, it is very important that these actions be contained in the stomach. Lots of people suffer from heartburn, which is the leakage of stomach acid into the lower esophagus. The acid causes an uncomfortable burning sensation, which sells very large numbers of antacids each year. Too much acid or infection by the bacterium *Heliobacter pylori* can cause a break in the lining of the stomach, which can turn into an ulcer. An ulcer can grow to become a life-threatening condition. This is clearly an example of a process that needs to be contained on the inside of the stomach.

In pregnancy, the developing fetus is protected by the uterine wall and nourished by food and oxygen provided via the placenta (Knöfler et al., 2019). The fertilized egg starts dividing and then forms a ball of cells that involutes onto itself. There is a definite inside and outside to the growing embryo. The position of cells early on defines their later developmental pathway. One recent discovery is that the uterus, long assumed to be sterile, is not (Wassenaar and Panigrahi, 2014). The source of infection is assumed to the primarily the female reproductive track. The placenta is thought to be a much less likely source. The organisms found in the amniotic fluid are *Ureaplasma urealyticum*, *Mycoplasma hominis*, *Fusobacterium nucleatum*, *Gardnerella vaginalis*, and *Bacteroides* spp. *Streptococcus* (Group B), *Escherichia coli*, and *Listeria monocytogenes* are sometimes found.

The Ultimate Inside

Life has spread to nearly every environment on Earth, perhaps with the top of Mount Everest and the South Pole. Living organisms can be found at more than 50 kilometers in the atmosphere in the stratopause. Organisms live in the deepest parts of the oceanic trenches known as the hadal or hadopelagic zone. Life can be found deep in the Earth. In addition, molecules associated with life (e.g., hydrogen and methane) are associated with deposits of serpentine, a semimetamorphic mineral that is squeezed up from great depths (Plümper et al., 2017; Vitale Brovarone et al., 2020). All of these together are contained in a thin layer of material that is approximately 60 kilometers thick. That is amazingly thin when compared to the Earth's diameter of 12,742 kilometers. Yet every living thing on Earth lives within that thin membrane.

Even within that thin membrane, life has had to survive constant threats. Some organisms and species survive, and other do not. During at least five periods, life on Earth has endured a mass extinction in which perhaps 90% of all life died. But life is resilient, and each time it has bounced back. We and all other living things on Earth can trace our lineage back directly to the beginning of life over 3.5 billion years ago.

We have survived the dramatic mass extinctions and the slow chronic extinctions. We may be in a sixth mass extinction, given the number and rate of species that have gone extinct. The current threat is different. In just the last few hundred years, humans have degraded the thin membrane to an exceptional extent. Climate change is accelerating. Weather is more extreme, and storms are more fierce. The ice sheets at the poles are melting. Plants and animals will have to try to adapt to these new norms. At some point, a threshold will be passed, and positive feedback systems will become unstoppable. Temperatures on Earth may run away. This is thought to have happened on Venus, where the surface temperatures are always in the hundreds of degrees. The most significant difference is that the previous mass extinctions were natural events. This new possible one is entirely of human making. Our activities threaten many species and perhaps even human life. At some point, we will have distinguished ourselves over all of the previous mass extinctions. Instead of the mass extinction changing the course of evolution, we humans will have ended evolution forever. The thin membrane can only take so much, and there is no other inside to protect us. There are only outsides.

7

Inside and Outside of Us

Humans also have an inside and an outside, except that the inside may not be quite what most people think it is. We typically look on our skin as our outside and everything else as the inside. But there's more to us than that. To mathematicians, humans are an elongated torus. The simplest way to envision a torus is to think about a donut. The "hole" in the human donut begins with the mouth and ends at the anus. The lumen of the gastrointestinal system is topologically outside of us. The inside includes the organs, muscles, bones, blood vessels, and other tissues. Other parts might seem like the inside, but are actually outside. These include the respiratory tract, nasal sinuses, urinary tract, female reproductive tract, and inside the ears. All of these are exposed to the outside environment to a greater or lesser extent.

Thus, the outside of humans is obvious and not so obvious. The obvious outside is our skin. The less obvious outside is the lining of our intestines. Both are interfaces with the external world, and both are exposed to the environment, the cells must be continually replaced by new cells. For example, the intestinal lining is replaced every 5–7 days. Those replacement cells come from stem cells that divide and differentiate into the terminal cell types.

In this chapter, we will take a look primarily at the topological outside of humans.

Outside: Skin

Our skin separates the inside from the outside. We have a lot of skin. It is the largest organ in our bodies. Skin has many critical functions (Dąbrowska et al., 2018). Our skin has a surface area of some 2 square meters and accounts for about 15% of total body mass (Rauma et al., 2013). It protects us from pathogens, ultraviolet light, chemicals, and injuries, and it preserves the water content of our body. Because our skin interfaces with the environment, it is subject to a lot of wear and tear. And we pay a lot of attention to it. In 2021, Americans spend $17.9 billion on skin care products (GlobalData, 2022). Sometimes the wear

The Biology of Us. Gary C. Howard, Oxford University Press. © Oxford University Press 2024.
DOI: 10.1093/oso/9780197664797.003.0007

and tear is more than simply cosmetic. We also spend about $85 billion each year on skin and subcutaneous diseases (Dieleman et al., 2020).

Composition

Skin contains several layers and multiple types of cells (Yousef et al., 2023) (Table 7.1). The skin is thickest on the palms of the hands and the soles of the feet. They both have an extra layer called the stratum lucidum. The three layers of skin serve different functions. The epidermis contains melanocytes that produce pigment and dendritic cells that aid the immune system. It also contains keratinocytes that produce vitamin D and Merkle's cells that are sensitive to touch and are connected to nerve endings. The outermost layer of the epidermis has 20–30 layers of cells. It's mostly keratin and dead keratinocytes. The dermis contains nerve endings, oil and sweat glands, and hair follicles. The hypodermis or subcutaneous layer contains fat, connective tissue and blood vessels. The skin is affected by the environment. For example, calluses make the layer thicker, and sunlight causes melanocytes to produce melanin that darkens our skin.

Table 7.1. Human Skin Layers

Layer	Sublayer	Layers	Contains
Epidermis	Stratum corneum	20–30	Keratin and horny scales of dead keratinocytes
	Stratum lucidum	2–3	Found in the palms and soles
	Stratum granulosum	3–5	Diamond-shaped cells with keratin precursors
	Stratum spinosum	8–10	Irregular, polyhedral cells with processes, dendritic cells
	Stratum basale		Basal lamina separates this from the dermis; cells are cuboidal to columnar stem cells that produce keratinocytes; melanocytes
Dermis	Papillary		Loose connective tissue
	Reticular		Fewer cells, connective tissue and collagen fibers, sweat glands, hair, hair follicles, muscles, sensory neurons, blood vessels
Hypodermis			

The greatest variety of cells is found in the epidermis, the outermost skin layer. These include keratinocytes, melanocytes, Langerhans cells, and Merkel cells. Keratinocytes are the most common. They produce keratin and lipids and thus are responsible for the water barrier of the skin. They also use sunlight to produce vitamin D. Melanocytes, as the name suggests, produce melanin, which is the pigment that gives skin its color. Melanin granules are transferred from cell to cell to spread the color throughout the skin. Langerhans cells function with the immune system. They contain MHC molecules, take up antigens, and transport them to the lymph nodes. Merkel cells are mechanoreceptors for gentle touches and interact with nerve endings. They are most common in the fingertips.

Obviously, the cells of our skin hold together to form a complete sheet that encloses our body in a relatively watertight manner. Those cells in the skin and the intestinal epithelia are held together by structures called cell junctions. There are three types: tight junctions, adherens junctions, and desmosomes (Garcia et al., 2018). In addition to holding the cells together, they regulate both diffusion of ions, microbes, and other material across the tissue and cell proliferation and migration. The tight junction surrounds a cell and attaches it to its neighboring cells by forming a protein seal between them. The two main proteins are claudin and occluding. They bind to cytoplasmic adaptor proteins, and those bind, in turn, to the actin cytoskeleton. The adherens junction involved the cadherin proteins (e.g., E-cadherin) that are embedded into the cell membrane. It binds to adaptor proteins (e.g., β-catenin and α-catenin) that are connected to the actin cytoskeleton. The desmosomes also include transmembrane and adaptor proteins. The transmembrane proteins (e.g., desmocollin and desmoglein) join to adaptor proteins (e.g., plakoglobin, plakophilin, desmoplakin). They, in turn, are bound to the cytokeratin intermediate filaments.

Function

Skin is amazing. It both separates us from and connects us to the outside world. It provides a barrier to protect our muscles and bones from infectious agents, but it also provides sensory information from the outside world through touch. Skin has several other important functions. It regulates body temperature, stores water and fat, prevents water loss, prevents bacteria and other pathogens from entering the body, and makes vitamin D. Vitamin D is a fat-soluble vitamin that aids in the absorption of important minerals, such as calcium, magnesium, and phosphate. Deficiency of this vitamin is a common problem. It is ordinarily synthesized in the epidermis when the skin is exposed to sunlight. However, many people in the developed world spend much less time in the sun and when we do, we use sunscreen. Thus, many people are deficient for this vitamin. Some fatty fish are a good source of vitamin D, but the biggest source in the United States is milk that is fortified with vitamin D. Severe deficiency results in rickets, but many elderly people also suffer from bone weakness due to vitamin D deficiency.

Evolution

Mammalian skin has changed very little over the last 210 million years (Maderson, 2003). It evolved to provide mammals with heat insulation, camouflage, and sensory input. For example, cats use their long whiskers to sense objects as they move around in the dark. One of the prominent features of skin is hair. Hair is thought to have evolved about 200 million years ago. It is found on all of the skin except for that on the soles of the hands and feet. Hair is made up primarily of the protein α-keratin.

Each hair has two parts. The follicle is found in the dermis of the skin and contains the stem cells that allow the hair to continue to grow. The shaft is the part that we see above the surface of the skin. The middle of each shaft is the medulla, which is unstructured. That is surrounded by a much more structured cortex. The cortex also contains the melanin that gives it color. The outer covering, or cuticle, is made of lipids. The shape of the follicle determines the characteristics of the hair. A round opening yields straight hair. An oval opening yields wavy or curly hair.

So why do humans have so little hair? Some scientists speculate that it correlates with when humans began to eat meat. The activity of hunting required them to be able to better control their body temperature. Humans lost hair and gained the ability to sweat (Kamberov et al., 2008). Sweating is essential to maintaining human body temperature. Sweat is released from those glands, and its evaporation cools the body.

Ecosystem

Because the skin interfaces with the external world, it provides a matrix for other organisms, such as bacteria, viruses, fungi, mites, and other creatures (Grice and Segre, 2011; Swaney and Kalan, 2021). Most are harmless. Some can be problematic.

After birth, the skin itself changes in function, structure, and composition for some time. For example, changes occur in the skin barrier and water-handling. As the infant grows, the skin expands and experiences high rates of cell proliferation. In like manner, the microorganisms on the skin of the infant change (Capone et al., 2011). During the early period, the skin is dominated by *Staphylococci*, but *Firmicutes* soon take over. This time is very important for the development of a healthy skin microbiome.

The microorganisms on the skin depend on the physical characteristics (e.g., moist, dry, sebaceous) of the skin (Byrd et al., 2018). The bends of the elbows and the feet are moist, and the dominant species are *Staphylococcus* and *Corynebacterium*. The most prevalent bacteria on sebaceous sites are *Propionibacterium* spp. Fungi are less common everywhere on the body. One group of fungi (*Malassezia* spp.) was found on the body core and arms. Feet show *Malassezia* spp., *Aspergillus* spp., *Cryptococcus* spp., *Rhodotorula* spp., *Epicoccum* spp., and more. One fungus, Candida (*Candida albicans*), lives in dark, warm, moist places, such as the folds of skin and the genital area. It also grows on the tongue, throat, or inside cheeks. Also called thrush or *candidiasis*, it's not common, but usually affects those who are immunocompromised (e.g., by AIDS or chemotherapy). Saheb Kashaf et al. (2022) combined classic bacterial cultivation techniques with metagenomic sequencing to assemble the Skin Microbial Genome Collection. It includes 622 prokaryotic species with 174 new bacteria species and 12 new genera. They also found 12 new eukaryotic species and thousands of viral sequences.

Skin Disorders Minor and Major

Being exposed to the environment, the skin is subjected to many hazards. Some are relatively trivial and inconvenient, but others are much more serious.

One of the most common problems with skin is a simple itch. For most itches, a good scratch is the best remedy. However, some itches do not stop and can be maddening. The mechanisms underlying an itch are still poorly understood. Oetjen et al. (2017) found specific proteins that are involved in itching. For example, a group of small proteins called type 2 cytokines activate sensory neurons to cause an itch. Chronic itch involves two other proteins, IL-4Ra and JAK1. JAK inhibitors offer some relief to sufferers. The IL-33 receptor, ST2, has been implicated in the itch of poison ivy (Liu et al., 2016). Poison ivy, poison sumac, and poison oak all contain the same oil, urushiol, that causes an allergic reaction on the skin. The good news is that this IL-33/ST2 pathway may point to new therapies to stop the terrible itching in poison ivy.

Itching is usually a minor irritation. However, about 20% of people suffer from a chronic itch, which is one defined as lasting for more than 6 weeks (Weisshaar, 2016). Chronic itch is associated with eczema, hives, psoriasis, conditions with social stigma (e.g., scabies, body lice), and more serious conditions (e.g., chronic kidney disease, liver failure, and lymphoma). Interestingly, itches result from itch-specific receptors that transmit signals from the peripheral nervous system to the spinal cord. Inflammatory cytokines (e.g., interleukins 4, 13, 31, and 33, oncostatin M, and thymic stromal lymphopoietin) serve as neurotransmitters to promote itching (Oetjen et al., 2017). The discovery of those receptors enabled the development of therapies for chronic itch. In fact, the study of itching provided insights into many other sensory pathways.

Too much sun can result in a painful experience. Sunburn is actually due to the ultraviolet radiation from the sun or a sunlamp. Several factors determine the risk of sunburn, including the length of exposure to the sun, cloudiness, proximity to the equator, and skin type. As a general rule, the lighter the skin, the greater the risk of sunburn (Sharma and Patel, 2023). Melanin, the skin pigment, protects against sunburn. However, even dark-skinned people can be sunburned and need to take care.

Exposure to the sun's ultraviolet B rays (Guerra and Crane, 2023) damages the cell's DNA by causing the formation of thymine-thymine dimers. The damaged DNA induces apoptosis in the cells and the release of inflammatory factors (prostaglandins, reactive oxygen species, and bradykinin). These are followed by the typical sunburn symptoms of vasodilation, edema, and pain. Sunburn is usually a painful, but minor problem, but it can have much more serious after-affects. A nice suntan can be attractive, but too much sun causes skin to age, and the risk of skin cancer is greatly increased. It is important to use sunscreen, wear hats, and avoid overexposure to the sun.

Skin Cancers

Skin cancers are the most common malignancies in the developed world. Age is a major risk factor for skin cancers, and since the elderly population is growing in the developed world, skin cancers are becoming more common (Guy et al., 2015). The best defense against skin tumors is to avoid excess exposure to the sun and to be examined by a physician or dermatologist regularly. This is especially important for older individuals.

The most common nonmelanoma skin cancers are squamous cell carcinoma and basal cell carcinoma (Sinikumpu et al., 2022). Squamous cell carcinomas may feature red patches, open sores, raised growths with a depressed center, or a wart-like appearance. They sometimes itch or bleed or crust over. Basal cell carcinomas are the most common skin cancers. Typically, they grow very slowly. They manifest with various features (e.g., open sores, red patches, pink growths, shiny bumps, scars, or growths with slightly elevated edges and a central indentation), but they are different in different individuals. These skin cancers rarely metastasize, and the majority of both types are resolved by surgery.

Age is risk factor for a very common precancerous state called actinic keratosis (Marques and Chen, 2023). These form rough, scaly patches and are caused by exposure to the sun. They develop mostly on areas that are exposed to the sun (e.g., face, lips, ears, back of hands, forearms, scalp, neck). Other risk factors include male gender, sun exposure, proximity to the equator, and immunocompromise. Actinic keratosis lesions are unsightly, but their real danger is that they sometimes turn into squamous cell carcinomas. The lesions are usually removed by freezing, chemotherapy, or surgery.

Melanomas account for only about 3% of all skin cancers, but they are, by far, the most dangerous (Shain and Bastian, 2016). Melanomas arise from melanocytes and so are typically dark brown or black but not always. Most are caused by exposure to the sun, but again not all. Their origin predicts some of their growth. Those without sun exposure come from benign or dysplastic naevi (moles). The sun-induced cancers arise in situ and may or may not follow a "typical" set of phases. A series of genetic mutations cause the melanocytes to lose their normal constraints on growth and become malignant. With additional growth of the tumor, later cells have more chances to experience additional mutations that cause the tumor to progress to later stages. Melanomas can develop to a stage in which they begin to metastasize. Cells from the tumor break off and travel to other sites in the body, where they can seed new tumors.

The Other Outside: The Digestive Tract

Mouth

The tissues of the mouth are continually exposed to foreign material as we eat, drink, and breathe, all of which are essential to life. The challenges include different foods, chemicals (e.g., alcohol, drugs), polluted air (e.g., smoke from fires or cigarettes), and infectious agents.

A mucus membrane provides a soft wet lining to protect the underlying tissues (Brizuela and Winters, 2023). It has three layers. At the top is a layer of squamous stratified epithelium that varies in its thickness and amount of keratinization, depending on the location in the mouth. Below that are two layers of connective tissue: the lamina propria and the submucosa, which is not present in every part of the mouth. The tissues are also different, depending on the area of the mouth. The mucosa on the movable parts of the mouth have nonkeratinized squamous epithelia and are much more flexible than other areas. Those parts that are connected to the nonmoving parts are keratinized to help absorb some of the stresses of chewing. The covering of the tongue has both kinds, since it is both mobile and involved in chewing.

Overall, the epithelia, like the skin, has four layers (Brizuela and Winters, 2023). It begins at the basal level where the stem cells reside. Those cells divide to produce new squamous cells, but also to maintain the balance between new cells and stem cells. The cuboidal cells move upward, and some become keratinized, depending on their ultimate function, and others do not. They lose their nuclei and other organelles. The process is similar to that of skin. The oral mucosa also contains melanocytes, Langerhans cells, and Merkel cells.

The oral mucosa has several important functions. It protects the tissues of the mouth with its layer of epithelial cells. It forms a water-tight barrier with cell–cell junctions, and dendritic cells and helper T cells (both immune cells) respond to foreign materials. The oral mucosa also secretes saliva. Saliva is a critical fluid (Humphrey and Williamson, 2001). This clear, slightly acidic secretion is a mixture of fluids from the salivary glands. Those glands include the parotic glands (opposite the first molars) and the submandibular and sublingual glands (in the floor of the mouth). Minor glands also produce saliva from the lower lip, tongue, palate, cheeks, and pharynx. The major glands produce more saliva than the minor glands, but the quality of the saliva differs from the different glands. Each day, we produce 1.0–1.5 liters of saliva. Those secretions are similar to serum (mainly from the parotid gland), mucous (minor glands), or mixed. The oral mucosa has fewer sebaceous glands than the lips. They secrete sebum. This fatty substance has been suggested to be involved in immunity, but its function is unknown.

Stomach

The stomach is a wonder. It secretes fluids that digest the foods we eat. If those fluids leak out of the stomach, they cause serious damage to other tissues. However, the stomach does not digest itself. That's a remarkable feat considering that the fluid contains hydrochloric acid. Our food also contains material that could damage the stomach, including alcohol and spicy foods. The stomach lining is not impervious to damage by the acid and other material. In fact, it is regularly injured, but recovers from that injury. The stomach is a hostile environment for living organisms due to the acid, lactoferrin, and immunoglobulins (Wallace, 2008).

The stomach has several layers of defenses. The lining cells secrete mucus that helps to protect the stomach from mechanical damage and also acts to trap microorganisms. It also provides a layer of neutral pH that separates the lining from the acid environment. One unresolved problem is how the acid gets from the lining, where it is secreted, to the surface of the mucus. One solution suggests that the stomach lining is hydrophobic, and so, the acid will not migrate back to it. In addition, the apical membrane of the lining seems to be exceptionally resistant to acid. The cells of the lining also turn over about every 2–4 days, which is much faster than the epidermal cells of other tissues. Even when there is damage to the lining, it is repaired rapidly. New cells migrate from healthy areas to fill in the gap from the damage. Prostaglandins are an important factor in the protection of the stomach lining. Aspirin and other nonsteroidal anti-inflammatory drugs hinder the activity of prostaglandins and can affect the stomach. In addition, the bacteria *Heliobacter pylori* is involved in many ulcers.

For some time, *H. pylori* has been associated with ulcers in humans. However, its relationship with humans is not completely clear. Over the last 100 years, the prevalence of that bacterium in humans in the developed world has become much less than in the developing world (Blaser, 2005). At the same time, the incidence of esophageal diseases has increased. Could *H. pylori* be protective against those diseases? So far, the answer is not clear.

Intestinal Epithelia

After food leaves the stomach, it enters the small intestines. The lining of the intestines differs from the more complicated epithelia of the skin. This simple intestinal epithelia is just one cell layer thick, but it performs some sophisticated functions, including protecting us from microbes and absorbing nutrients and water.

The lining comprises columnar epithelial cells, but the structure of the lining is quite different from that of the skin. The surface has many, many finger-like projections called villi, and the columnar epithelial cells of the villi are covered with microvilli called a brush border. These serve to enormously increase the surface area that is available for absorption of nutrients. In the depths between the villi are the crypts, which contain the stem cells that produce new cells that replace the worn-out cells of the villi. For many years, most scientists have believed that a second pool of stem cells existed. When the first pool of stem cells was exhausted, the second set would be activated to produce new cells. Intriguingly, Jadhav et al. (2017) found that, in mice, when the first set is depleted, mature epithelial cells dedifferentiate to become stem cells to provide the new cells. This process likely occurs also in humans.

The True Inside

Our outside and inside are not so obvious. The obvious outside is our skin. The less obvious outside is the lining of our intestines. Both are our interface with the external world, and both are exposed to the environment. Our true inside is contained between the skin and the mucus membranes and contains our organs, bones, muscles, glands, and other tissues. Even the raw numbers are amazing. We have more than 600 muscles. Our circulatory system of arteries, veins, and capillaries is more than 96,000 kilometers (60,000 miles). We have over 200 individual bones that are connected by tendons, ligaments, and cartilage. Infants have a few more that fuse as they grow. We are about 50% water by weight. We process about 200 liters of blood each day to remove about 2 liters of waste. We excrete about 1.4 liters of urine each day. Our brain contains about 100 billion cells. All those systems have to work together correctly for us to remain healthy. And to do that, they must develop as intended and be held in place.

Those activities that need to be "inside" are contained inside specific spaces. Blood is contained within vessels. Stomach acid stays in the stomach. A growing fetus is held within the uterus. Breaches in those containments result in trouble.

8

Food and Eating

"Crawling at your feet," said the Gnat (Alice drew her feet back in
some alarm), "you may observe a Bread-and-butter-fly. Its wings are
thin slices of bread-and-butter, its body is a crust, and its head is a
lump of sugar."

"And what does it live on?"

"Weak tea with cream in it."

A new difficulty came into Alice's head. "Supposing it couldn't
find any?" she suggested.

"Then it would die, of course."

"But that must happen very often," Alice remarked thoughtfully.

"It always happens," said the Gnat.

—Lewis Carroll (1909),
Through the Looking Glass: And What Alice Found There

The Gnat was right. Eating is critical for every animal. Sometimes we eat,
and sometimes we are eaten. Plants are producers. They tap the energy of
the sun, the ultimate source of essentially all energy used by living organisms
on Earth. They store it in carbon-based compounds made from the carbon
dioxide they absorb from the atmosphere during photosynthesis. Animals
and humans eat the plants, and animals and humans eat other animals. We
humans eat a lot, and essentially everything we eat is the product of other
living organisms.

In the overall food chain, we humans are the ultimate consumers. We obtain energy by eating plants or eating other animals that eat plants. In addition to energy, we obtain other benefits from food. We produce several of
the 20 amino acids that are the building blocks of proteins, but we get others
from our food. Those are called essential amino acids (i.e., histidine, isoleu-
cine, leucine, lysine, methionine, phenylalanine, threonine, tryptophan, and
valine) (Lopez and Mohiuddin, 2021). In like manner, we make many of the

The Biology of Us. Gary C. Howard, Oxford University Press. © Oxford University Press 2024.
DOI: 10.1093/oso/9780197664797.003.0008

fatty acids that make up the lipids we need to build membranes and make hormones (Delage et al., 2019). Others (e.g., linoleic acid, α-linolenic acid) we have to obtain from our food. We also get vitamins, minerals, and other nutrients that catalyze biochemical reactions that are critical in normal cell function, growth, and development. The vitamins are organic molecules and include A, C, D, E, and K, and thiamin (B_1), riboflavin (B_2), niacin (B_3), pantothenic acid (B_5), pyridoxal (B_6), cobalamin (B_{12}), biotin, and folate. Minerals are inorganic elements and include calcium, phosphorus, potassium, sodium, chloride, magnesium, iron, zinc, iodine, sulfur, cobalt, copper, fluoride, manganese, and selenium. Calcium and phosphorous are important in bones and various reactions. Iron is used to make red blood cells to transport oxygen. Finally, we get water from our food and drink.

In Chapter 4, we discussed the living organisms that humans use for food, and we toured a supermarket to learn about those life forms. Here we will look at how eating has influenced human evolution and how our anatomy and physiology have evolved with regard to food and how we prepare it.

Food can tell us a lot about humans. What we eat has changed over our history, and the way we eat it has too. Some things we take for granted today were astounding technological breakthroughs in their day. Today's kitchens are well stocked with foods and utensils. Refrigerators, stoves, ovens, and microwaves are common. To prepare a meal today, we might open a package of rice or pasta and add it to a pan of water heating on the stove. Who gives a second thought to turning on the stove, oven, or microwave? But cooking with fire greatly increased the nutrients that humans could extract from food. Likewise, the use of containers made cooking more efficient, and agriculture provided a larger and more dependable supply of food. Throughout all of this, the living organisms that we use for food changed. Each of these changes was a dramatic step forward in civilization.

Of course, eating is only the beginning of a very complicated and involved process. The food needs to be broken down and moved to where it is needed, and structures and mechanisms have evolved to facilitate those processes. Teeth, the tongue, and salivary glands begin the process of digestion. The ability to swallow and maintain breathing is needed. The stomach, intestines, and gall bladder play their roles in digestion. The blood stream moves nutrients to where they are needed. The pancreas and islets manage sugar levels, and the liver aids in the transport and processing of lipids (fats). Finally, the liver, kidneys, and colon handle the disposition of waste products. At a molecular level, the process is far more complicated.

Box 8.1 How Long Can a Person Live without Food?

> Whilst it is true that man cannot live without air and water, the
> thing that nourishes the body is food. Hence the saying, food is life.
> —Mohandas Gandhi (1959, p. 177)

On February 10, 1943, Mahatma Gandhi, age 74, began a hunger strike that lasted for 21 days. Gandhi had turned hunger strikes into a weapon to resist continued British rule of India. He subsisted on only small sips of water.

Humans can survive a surprising length of time without food. Cases of starvation of a month or more are well documented (Lieberson, 2004). It is difficult to predict how long a person can survive without food or with very little food. Multiple factors, such as hydration and the amount of body fat, complicate predictions. The body has some ability to alter its metabolism to conserve energy. In cases of anorexia nervosa, patients typically die from organ failure (i.e., cardiac infarctions) when their body weight is 60–80 pounds. Patients with end-stage cancer often die when they have lost 35%–40% of their body weight. This generally takes 1–3 weeks.

The Evolution of Eating

Food we eat has had a significant influence on our anatomy and physiology, and in turn, humans have had a great effect on the foods that we eat.

Food wasn't always as convenient as it is now. For millions of years, hominids were hunter-gatherers, who foraged for edible plants and hunted or scavenged animals. Their diet was similar to that of the other great apes. Our early hominid ancestors likely ate a diet of many plant foods, and they ate it raw. Most large apes still subsist on a plant-based diet, and it is reasonable to assume that our early ancestors did the same. Australopithecines had large molars with thick crowns, which are consistent with a plant-rich diet.

Diet is a critical element in evolution. The foods hominids consumed and the physiological mechanisms to digest them evolved together. For example, 2.3–4.4 million years ago, early australopithecines (i.e., *Australopithecus anamensis*, *A. afarensis*, *A. africanus*) underwent dramatic changes that greatly improved their lives. By reviewing evidence involving tooth size and shape, enamel structure, dental microwear, and jaw mechanics, Teaford and Ungar (2000) found that, over time, these early hominids began to eat harder, more abrasive foods. Those changes were accompanied by changes in their teeth. The australopithecines had relatively small incisors and large, flat molars with thick enamel. Unlike their great ape relatives, they would have had trouble eating large fruits with thick husks, but they could have easily dealt with soft fruits and leaves and crushed hard abrasive materials. Thus, they might have had an advantage of a wider variety of foods than the other groups, and that advantage might have allowed them more options in varied habitats.

Human diets had two major changes: eating meat and cooking. Meat is not essential, but it does provide concentrated protein and fat. However, vegetarians who cook their food live very well. Our hunter-gatherer ancestors survived on raw foods, but they had to spend a lot of time finding and eating food. Their food supplies were also seasonable. Those today who eat only raw food have a hard time getting sufficient calories. Some research has shown that they have low body-mass indices or BMIs and even have interrupted ovulations. Yet, their diet includes domesticated foods that are more digestible than those available to our hunter-gatherer ancestors, and even those items are typically prepared with food processors.

One line of evidence for the difference in diets comes from studies of the isotopes of carbon and nitrogen in the collagen of animals and the two groups. (Isotopes are atoms of the same element that have different numbers of neutrons. For example, nitrogen has two naturally occurring isotopes that have either 14 or 15 neutrons). The ratios of nitrogen 15:14 are well known for the Holocene, and those ratios have been used to identify carnivores, omnivores, and herbivores. The ratio was lowest in plants, higher in herbivores and higher still in omnivores, and highest in carnivores. They are even higher for top aquatic consumers. Those for omnivores also vary depending on the mixture of plants and animals they are eating. The ratios of carbon 13:12 also vary with diet among aquatic and terrestrial consumers.

The Neanderthal diet was varied. Richards and Trinkaus (2009) used those ratios to compare diets of Neanderthals and early modern humans. They found that the Neanderthal diet consisted mainly of large herbivores for a long time (about 37,000 to 120,000 years ago). However, the diet of early modern humans was much more variable and even included both marine and freshwater foods. They conclude that the ability to obtain food from so many more sources might have given modern humans an advantage over the Neanderthals. The isotope analysis is limited to only food from meat.

Neanderthals might have also eaten plant foods. Henry et al. (2014) looked at the plants consumed by the two groups by looking at microremains, such as starch grains and phytoliths, in the dental calculus and on tools. They found that both Neanderthals and modern humans ate a wide variety of plant foods. Weyrich et al. (2017) examined the genomic information preserved in the dental plaque of Neanderthals at various locations. Some ate primarily meat. Others ate no eat, but lived on mushrooms, pine nuts, and moss. Sistiaga et al. (2014) examined the fecal remains from Neanderthals from about 50,000 years ago in Spain by gas chromatography and mass spectrometry. They found levels of a cholesterol metabolite that confirms that Neanderthals ate predominantly meat. However, they also found 5b-stigmastanol, which indicates significant plant intake.

Cooking food was a major advance for humans, but when humans actually began to use fire is not clear. Certainly, by the Upper Paleolithic (about 50,000 years ago), fire was commonly used to prepare food. The evidence for its controlled use in the Lower Paleolithic (300,000 to 1.5 million years ago) is more controversial. Early hominins (*Homo habilis*, *H. erectus*, and *H. ergaster*) may have begun to use fire as early as the Lower Paleolithic (Wrangham, 2017). What is not clear is whether they could start fire or simply used available fire. Neanderthals seem to have used fire for cooking, but it is also unclear whether they actually knew how to make fire or if they simply used it when it was available (Henry, 2017).

Wrangham et al. (1999) suggest that anatomical changes to early *H. erectus* (e.g., smaller teeth) indicate that humans were cooking about 1.9 million years ago. They also note that female body mass increased at that same time. From these observations, they conclude that cooking food had significant effects on human anatomy and sociology. Teeth are an important factor in evolution (Smith, 2013). The teeth of early hominins was more similar to apes than Neanderthals or *H. sapiens*. Neanderthals and modern humans coexisted in Europe before the Neanderthals were replaced by modern humans. Why isn't clear, but success in obtaining food is a possibility. Modern humans used sexual division of labor and more complex projectile weapons. These allowed a better diet than that of the Neanderthals that allowed them to outcompete the Neanderthals.

A seeming paradox has been found in early humans. *H. erectus* had a larger brain and body than its predecessors, but also smaller teeth and a weaker bite and smaller gut. How could this be? The usual explanation is that humans began to eat meat, processed food with tools, or started cooking. Problems exist for each of these explanations. Food preparation began with the processing of foods with stone tools. Zink and Lieberman (2016) looked at that topic. They assumed that meat accounted for about one-third of the diet. Under that assumption, they estimated that processing food reduce the number of chewing cycles by about two million per person per year, and the needed mouth force would have declined by 15%. These findings argue for the importance of using stone tools to grind tubers and grain and to slice meat.

About 10,000 years ago, humans began to leave their hunter-gatherer lifestyle and settle down (Gibbons, 2014). We started to develop agriculture and to domesticate animals, build communities, and specialize labor. The first grains to be domesticated (e.g., sorghum, barley, wheat, maize, and rice) provided a more dependable food supply than foraging. The increased nutrition can be seen in the increased rate of childbirth (2.5 years for farmers vs. 3.5 years for foragers). Swidden farming is one step in the evolution of agriculture. It involves the temporary clearing of areas of forest for growing crops. It is sometimes called "slash and burn" farming, but that might put it into a less favorable light.

Humans have not just been passive consumers of various foods. They have also taken over the evolution of specific plants and animals as food crops. The domestication of plants and animals began in Mesopotamia about 10,000 years ago. Those plants included wheat, barley, lentils, and peas. Other areas started with different plants, such as rice in Asia and potatoes in the Andes. Goats and sheep were likely the first animals to be domesticated (Ahmad et al., 2020). Almost from the beginning, humans began to select for desirable characteristics in plants and animals. Modern molecular genetics has provided a powerful tool to examine the origins of domesticated animals (Frantz et al., 2020).

The origins of maize or corn (*Zea mays mays*) are controversial, but humans had a great deal to do with it. The plant is thought to have originated from teosinte (*Z. mays parviglumis*), a wild grass, in southern Mexico about 9,000 years ago (Piperno et al., 2009). Humans selected for specific characteristics and transformed it from a simple grass with small grains to a major food crop with the large grains of today (Tian et al., 2009).

Like maize, the potatoes of today (*Solanum tuberosum* L) are nothing like they were before humans intervened in their evolution about 8–10,000 years ago. Potatoes originated from the wild *Solanum* species in the Andes of southern Peru. They are in the nightshade family and produce a glycoalkalois toxin solanine. Fortunately, it is present in only small amounts, except in the green-colored sprouting regions of the potato. They provided a major energy source for people in what is now Peru, Bolivia, and Ecuador and later spread throughout the other areas of South America. Today they are the one of the most important crops for human consumption. Domestication involved changes in multiple traits, including increased yield, reduced toxicity, and stress tolerance (Meyer and Purugganan, 2013). The vast majority of the potatoes available in North America derive from potatoes taken to Europe centuries ago. Today we have over 5,000 varieties of potatoes of various colors (e.g., blue, yellow, white, orange) to choose from. For example, the Russet Burbank, a very popular variety for baking and French fries, originated over 100 years ago. After the manipulations by growers, domesticated potatoes have a complex gene pool that combines the gene sequences from over 100 subspecies. Plants are polyploid and can add copies of gene fairly easily. This characteristic has added to the diversity of the potato genome. Hardigan et al. (2017) completed an extensive study of potato genomes that identified 2,622 genes under selection but only 14%–16% were shared by the North American and Andean cultivars. These results demonstrate the extent of the selection over the years. They also showed that the remaining wild *Solanum* species possess a great deal of additional genetic potential for further selection.

Cooking

Cooking seemed to begin only about 500,000 years ago. The origins of meat eating are less clear. Humans are listed as omnivores, but we could not survive on a diet of raw foods. In fact, humans have adapted themselves so well to a cooked diet that Furness and Bravo (2015) suggest that humans should be classified as cucinivores. The modern human anatomy and physiology have changed to accommodate a cooked and richer diet. For example, multiple structures (e.g., mouth, jaw muscles, jaw, teeth, stomach, and colon) are all much smaller than those of older ancestors. Our colon is shorter than those of other primates. Our teeth and jaws are smaller and less muscled. One example of a change in diet involves grains. At first, cooking consisted of roasting. Later humans began to use boiling. They heated rocks and dropped them into temporary containers.

Box 8.2 When We Are on the Menu

Food is important to us, but it also is important to other carnivores. Throughout human history, we have been the meal rather than the diner. Our ancestors and even some who live in parts of the world today were and are still considered food by large predators.

The numbers of deaths from wild animals are difficult to determine. From 1979 through 1990, 157 deaths from animal attacks on average occurred in the United States (Langley and Morrow, 1997). These included both venomous and nonvenomous causes. Conover (2019) reviewed published reports and estimated that 700 people were killed by wildlife annually in the United States. In either estimate, the numbers are relatively small, but certainly real. They must have been much greater among our ancestors, who were far more exposed to predators than we are today.

We can extrapolate what ate our ancestors by looking at what animals prey on primates today (Dunn, 2011). Leopards account for about 70% of baboon kills, and they kill many other monkeys chimps, and young gorillas. Lions, tigers, cougars, and jaguars also take primates. Wolves and other dogs also prey on primates. Hyenas have been known to kill people. Many large predators (e.g., sabertoothed cats) may have attacked early humans. Finally, primates eat primates, and so, hominids might have preyed on other hominids.

We can be a food source for very small organisms too. Mosquitos, fleas, ticks, and many other organisms enjoy our blood, skin, and more. An amazing list of internal parasites also feed off of us (NewScientist, 2009). Among them are hookworms (*Necator americanus*), roundworms (*Ascaris lumbricoides*), blood flukes (*Schistosoma mansoni*), tapeworm (*Taenia solium*), pinworms (*Enterobium vermicularis*), nematodes (*Wuchereria bancrofti*), and protozoans (*Toxoplasma gondii*, *Giardia lamblia*, *Entamoeba histolytica*). Finally, at death, we all become a feast for insects, microorganisms, and more.

Uncooked grains have little available nutrient value for humans. However, once they are cooked, they are more digestible so that more nutrition can be obtained from the same amount of food. For example, plant lipids are contained in oil bodies that have oleosin proteins on the surface. Cooking denatures the proteins and makes the lipids more available for digestion. Animal lipids are contained lipid droplets, which are also surrounded by proteins. Cooking increases the energy available in foods by about 30%. No human population today lives on only raw foods. However, studies of other primates have shown that even small increases in food energy have large effects on the animals' health.

In addition to fire, the use of containers for cooking was also an important step for humans. North Africa and East Asia are the two areas where the oldest pottery has been found. Dunne et al. (2016) examined traces of organic material left on pottery shards from Libya from 6,400–8,200 years ago. They found that grasses and aquatic plants were commonly cooked in those containers and that the containers seemed to show that more plant foods were processed than meat. Thus, even when early humans were settling down, plants still made up a significant portion of their diets.

Cooking is a key part of digestion. It kills microorganisms and parasites that might be in the food. It also begins the breakdown of the components of the food so that our digestive processes can more effectively extract the nutrients in the food. Several biochemical reactions are known (Vilgis, 2015). The Maillard reaction, named for French chemist Louis Camille Maillard, is one of the most important. When meat is cooked, its proteins and sugar molecules on the outer surface of the meat combine in a multitude of complex reactions to form new aromatic compounds that give roasted or fried meat it flavor. Carbonyl groups on the sugars in the meat react with amino groups on amino acids in the protein. The middle of the meat does not undergo the reaction. Cooking also affects the starches in foods, such as potatoes. The starch particles are burst and expanded by the steam that forms inside the food as it cooks. The starch expands 100-fold and becomes far more available for further digestion. In caramelization, the carbohydrates or sugars in food lose water during cooking. The sugars are isomerized and polymerized.

Many biological substances are polymers of small components (see Chapter 6). They have a compact structure that serves storage or other functional needs. For humans to use them for food, the items need to be processed physically or chemically to make their nutritional value more accessible. Some, such as grains, need to be ground or crushed. Others must be cooked. We can see some example of basic biology with some simple cooking techniques, such as boiling an egg and making popcorn.

Humans consume starch in the form of grains (wheat, barley, rice), potatoes, maize (corn), and cassava. Starches are the major energy-storage molecules in plants and are polymers of sugars. Glucose is water soluble and can interfere with the salt balance in the cells (osmotic pressure). Starch is not soluble and does not contribute to the osmotic pressure, and thus is a convenient form for storing energy. Amylose is a linear form (20%–25% of plant storage), and amylopectin has branches (75%–80%). Animals store carbohydrates in a highly branched form called glycogen.

The three-dimensional arrangement of how the molecules are connected is very important in starch and other polysaccharides. Some orientations are more stable than others. The alpha bonds in starch and glycogen are easy to break so the energy of the starch is readily available. Other polysaccharides are formed with the much stronger beta bonds. Thus, chitin, cellulose, and peptidoglycan are used for structures and protection.

The energy content of starches is much more accessible to humans when the starches are processed before eating. Grinding and cooking help to open up the starch grains. Grains are typically ground to make bread, and potatoes and corn are cooked. Breads are essentially ground grain and water that is baked. Those two ingredients form the most simple breads. Other additives change the final product. Bread can be leavened to make it rise and be more porous and soft. Leavening is done by chemicals (baking soda or baking powder) or yeast. Baking powder and baking soda both need an acid to activate them. Baking powder contains the basic sodium bicarbonate and an acid monocalcium phosphate that react to produce carbon dioxide that causes the dough to swell and spread. Baking soda is pure sodium bicarbonate and so relies on acids in the dough (e.g., brown sugar that contains molasses). Yeast are fungi. The most commonly used is *Saccharomyces cerevisiae*. They digest the sugars in the flour and release carbon dioxide that causes the dough to rise.

Popcorn is a special type of corn (or maize) that has a strong hull, a starch-filled endosperm, and 14%–20% water. As the popcorn is heated, the water turns to steam. The temperature in the kernel reaches about 180°C and the pressure about 135 psi. The pressure inside exceeds the hull's capacity to contain it, the kernel pops and the starch expands 20–50 times its original volume. The steam also causes the starch and proteins in the kernel to gelatinize and soften. As the kernel rapidly cools, those components form a sort of foam that is now digestible to humans.

A similar process occurs when potatoes are cooked. For example, when potatoes are baked, the starch granules begin to absorb some of the surrounding water. As the internal temperature of the potato increases, the water turns to steam and causes the granules to expand. The expansion can be quite violent if the thick potato skin is not pierced to allow the steam to escape. The potato will simply explode in the oven.

Meat is roughly 75% water, 20% protein, and 5% fat and other molecules. The protein molecules are condensed into coils. Cooking denatures the protein by breaking those bonds and allowing the protein to uncoil. Collage in the muscle gelatinizes and the fat melts.

Cooking an egg shows what happens to proteins when they are cooked. Proteins are long linear chains of amino acids. They coil up to form a three-dimensional structure that is specific to their sequences of amino acids. The structure of proteins and the importance of the structure to their function is covered in detail in Chapter 4. As a brief review, some amino acids are hydrophobic, some are hydrophilic, and others are neutral. The hydrophobic ones like to be close together, and so do the hydrophilic ones. These preferences form the basis for the three-dimensional folding that they assume. Egg white is about 12% protein, and the yolk is about 16% protein. In a raw egg, the proteins are in their native state and have their normal three-dimensional structure. When the water containing the egg is heated, the proteins in the egg begin to denature. That means that the normal three-dimensional structure is disrupted so that the proteins begin to clump together. As a result, the white and yolk harden. Sometimes the egg shell cracks and the white begins to leak out. If we add some table salt or vinegar to the water, the leakage will stop. The reason is that those substances also cause the protein to denature. So the leaking protein will clump and seal the crack. Hard-boiled eggs also sometimes discolor at the boundary of the white and the yolk. Iron ions in the yolk react with hydrogen sulfide gas in the white to form the green compound iron sulfide. The eggs are completely safe to eat, but they are less desirable. The reaction can be mitigated by moving the cooked eggs quickly to cold water.

Fats and oils are both lipids. The difference is that, at room temperature, oils are liquid, and fats are solid (Talbot, 2017). In meats, fats are off-white in color and solid in consistency. During cooking, they melt at 130°F–140°F. Not only do they melt, they also provide much of the flavor of meat.

Fruits and some other foods turn brown. Enzymes, such as polyphenol oxidase, catalyze a reaction that oxidizes the compounds found naturally in fruit. Those compounds are called phenols, and they are turned into quinones, which then become the brown pigments called melanins. The enzymatic browning reaction is usually bad for growers and grocers. Fruit with brown spots is less attractive to shoppers, and so this reaction is economically detrimental to fresh fruit, vegetables, and shrimp. The reaction can be stopped with acids or heat. For example, lemon juice lowers the pH so that the reaction is less efficient. Blanching destroys the enzymes. However, the same browning is good for some foods. The same reaction brings out the flavor in tea, figs, and raisins.

Digestion

Eating and digestion require a combination of body parts, enzymes, and activities and tremendous coordination of all of these.

Mouth-watering. The phrase conjures up great food in our minds. The smell or sight of food stimulates the brain to direct the salivary glands to produce saliva and begin the first step in digestion. Importantly, the stimulation of secretion of saliva prepares the mouth to receive food. In humans, the exact pathway of the stimulation is not completely understood (Pedersen et al., 2018). It's controlled by the autonomic nervous system, which controls many functions that we do not have to consciously think about (e.g., breathing, heartbeat, digestion). Those functions are controlled by the medulla oblongata in the brainstem, which connects the higher brain areas to the spinal cord. The salivatory nuclei in the medulla oblongata receive the input along with other impulses from higher brain areas and then turn on secretion.

Saliva has several functions. It lubricates food so that it can be more easily swallowed, but more importantly, it contains amylase. This enzyme breaks down starches. Starch is a critical component of the human diet. Without amylase, humans cannot break down starch into digestible sugars. Interestingly, the availability of amylase in saliva might have accelerated human evolution (Perry et al., 2007). Humans have many more copies of the amylase gene than chimpanzees. The additional copies might have given humans the ability to obtain more benefit from their food. Saliva has other functions too. It lubricates the tissues of the mouth, prevents the growth of bacteria, maintains a proper pH, and forms a coat on the teeth that prevents calcium loss. As the saliva flows from the glands and through the ducts toward the mouth, its composition changes. Most of the sodium is reabsorbed, and potassium and bicarbonate are added. Saliva also helps to keep the mouth clean. It washes away food debris, and it contains an enzyme called lysozyme that attacks bacteria. When we are asleep, the secretion of saliva decreases. The "morning breath" that we often experience results from the growth of bacteria in the mouth overnight.

Humans have three major pairs of salivary glands (i.e., parotid, sub-mandibular, and sublingual) and hundreds of smaller glands (Roblegg et al., 2019). A fourth pair has been proposed (i.e., tubarial) although some scientists disagree. All together they produce about 0.5–1.5 liters of saliva each day, and the production is controlled by the parasympathetic nervous system. They produce different types of saliva (Holmberg and Hoffman, 2015) The parotid secretes mostly alpha-amylase and serous fluid. The sub-mandibular and sublingual secrete a mix of alpha-amylase, mucus, and serous fluid. The tubarial secretes mostly mucus. Salivary glands are found in many mammalian and bird species. In some animals (e.g., snakes, Gila monsters), they have evolved to produce venoms. In others (e.g., insects, especially silk worms), they produce silk, which consists of protein fibers.

The first and most obvious structures involved in eating are teeth. The adult or permanent teeth in humans normally include 32 teeth embedded in the mandible (lower jaw) and maxilla (upper jaw). The roots of the teeth are covered by the gums. In the front are eight incisors to cut food. Behind the incisors on each side are a total of four canines, which are sometimes called vampire teeth, that tear food. Farther back are the premolars and molars that crush and grind food. There are eight premolars or bicuspids. Finally, 12 molars fill out the complement of adult teeth. The rear-most molars are also called wisdom teeth. They typically erupt at about age 20. The adult teeth are our second set of teeth. The primary set or baby teeth include only 20 with eight incisors, four incisors, no premolars, and eight molars. These develop in utero, but normally break through the gums in infants. They are eventually replaced by the permanent teeth between the ages of 6 and 12 years.

Teeth have four main parts. The outermost and hardest layer is enamel. Enamel is 96% mineral and is made of hydroxyapatite, a form of calcium phosphate. In contrast, dentin is a 70% organic matrix of collagen-type proteins and supports the enamel. The cementum is about half inorganic material and provides an attachment for ligaments that hold the tooth in place. The dental pulp is soft organic material that contains blood vessels, nerves, and other cellular components.

The first step in digestion is chewing. Chewing is part conscious and part unconscious, and it stimulates the production and secretion of saliva. Food is chewed as muscles cause the teeth to alternately open and close to grind or crush the food. Four muscles are primarily involved in chewing (Basit et al., 2021). The temporalis, medial pterygoid, and masseter muscles cause the lower jaw or mandible to close. The lateral pterygoid muscle with an assist by gravity depresses the lower jaw. The tongue and cheeks help to position the food for chewing to occur. These actions increase the surface area of the food and mix it with the saliva so that the salivary enzymes can begin to digest the carbohydrates. Chewing is something that herbivores do. Carnivores tend to swallow their food in chunks.

Once food is swallowed, it passes into the esophagus, a muscular tube that connects the back of the pharynx at the back of the mouth to the stomach (Squier and Kremer, 2001). Food and drink must pass through it to reach the stomach. We consciously begin the process of swallowing, but the involuntary nervous system quickly takes over at the top of the esophagus. Swallowing stretches the smooth muscles that make up the tube. This action stimulates intrinsic nerves to begin peristalsis, which causes the muscles to rhythmically contract in a wave from top to bottom in a way that moves food or liquid along. Muscular sphincters at the top and bottom of the esophagus provide one-way valves that let material go down, but not up.

The stomach performs the second step in digestion. This hollow organ lies just below the diaphragm and holds about a liter of food. It secretes digestive enzymes and acid to break down food. The main enzyme pepsin breaks proteins down into small pieces called peptides and individual amino acids. It is secreted as its inactive form called pepsinogen by gastric chief cells and converted into the active form of pepsin by the acidic conditions of the stomach. The gastric chief cells are found deep in the mucosal layer of the stomach. Gastric lipase breaks down lipids or fats. They are also secreted by the gastric chief cells and complete nearly one-third of the digestion of lipids.

Parietal cells in the stomach secrete hydrochloric acid. The acid denatures proteins to promote their degradation. It also kills any bacteria and viruses that might be in the food and as noted above activates pepsin. The acid is a key part of the digestive process, but the cells of the stomach itself are made of proteins, and the acid can erode them too. The stomach protects itself with a lining that is made of mucous cells that secrete mucin and bicarbonate. The parietal cells also make intrinsic factor. This glycoprotein (a protein with a sugar group attached) helps the intestines to absorb the critical vitamin B12 in the ileum. When we eat, the food causes the stomach to swell, and that action prompts G cells near the bottom of the stomach to secret the endocrine hormone gastrin into the bloodstream. It causes the stomach to produce acid and intrinsic factor.

The partially digested mass of food, digestive enzymes, and acid in our stomach is called chyme. We are all familiar with what it looks like. We have all on occasion thrown up this pulpy, acid soupy mass as vomit. Vomiting seems to be specific to species. Rats, horses, and cows rarely vomit. Those of us who have cats and dogs as pets know that they vomit only too often.

As disgusting as it is, vomiting is usually a carefully coordinated physiological action (Bowen, nd). Here are the events. We take a deep breath. The glottis closes and the upper esophageal sphincter opens. The soft palate closes off the posterior nares. Negative pressure is produced when the diaphragm contracts sharply, and this opens the lower sphincter. The muscles of the abdominal wall contract to squeeze the stomach and push its contents up though the esophagus and out the mouth. At other times, those actions are bypassed as a sudden ejaculation of the stomach contents pushes them up through the mouth. This is referred to as projectile vomiting. The clinical term for vomit is emesis.

Control of vomiting is in the brainstem. Inputs from various sources are coordinated by the bilateral vomition centers in the reticular formation of the medulla oblongata. Some of those inputs come the stomach (e.g., distention of the stomach). Others come from regions outside the stomach, including the peritoneum, heart, bile ducts, and other organs. For example, a gall stone in the bile duct can trigger vomiting. In addition, other areas of the brain can induce vomiting (e.g., motion sickness, odors, fear). The chemoreceptor trigger zone can sense problems (e.g., hypoxia, ketoacidosis, uremia) that might cause vomiting. Antiemetic drugs act in this area to suppress vomiting. Aside from the discomfort, a bout of vomiting does not usually cause serious problems. However, repeated vomiting can result in disruption of the acid-base balance, dehydration, and electrolyte depletion. Also aspiration of the vomit contents can cause aspiration pneumonia or even death if the person is unconscious and unable to clear the material from the airway.

Once food has been digested by the stomach, it passes into the intestines, where most of the absorption of nutrients takes place. The small intestines in humans are 3–5 meters (9.5–16 feet) long. However, when all of the folds, villi and microvilli are considered, the amount of surface area available for absorption is truly impressive: about 30 square meters. That's about 320 square feet, or the area of a large living room. That great length of tissue must be carefully packed into the abdomen. Savin et al. (2011) described the development of the gut and how the placement of the loops is carefully controlled.

The small intestines are divided into three sections. The duodenum is 20–25 cm and is involved mostly with further digestion of the chyme or the partially digested material from the stomach. It also receives enzymes from the pancreas and bile from the liver and produces bicarbonate to neutralize the acids from the stomach. Proteins are digested to amino acids. Lipids are digested to glycerol and fatty acids. Some carbohydrates are digested to simple sugars. Others, such as cellulose, are not and simply continue on to elimination. The jejunum is about 2.5 meters long and contains the folds, villi, and microvilli that accomplish most of the absorption of nutrients into the bloodstream. The last stretch is called the ileum. It is about 3 meters long and also contains folds and villi for absorption. However, it focuses on absorbing vitamin B and the bile salts and any remaining nutrients.

Our digestion is helped by more than 1,000 species of anaerobic bacteria that live in the intestines (e.g., *Bacteroidetes* and *Firmicutes*) (Zhang et al., 2015). These bacteria were described in detail in Chapter 5. They supply needed nutrients, make vitamin K (useful in clotting), and digest cellulose and other molecules that we cannot. Some digest dietary fiber and polyphenols. They also protect the lining of the intestines from antigens and pathogens.

The lining of the intestines is subject to a lot of wear and tear. The cells are damaged and need to be constantly repaired or replaced. An amazing process has evolved to accomplish this. In the crypts between the villi, stem cells divide to provide a steady supply of fresh cells that move up the villus to replace the worn-out cells, which are shed from the villus.

The large intestine is shorter but thicker than the small intestine. It consists of the cecum, colon, rectum, and anal canal. Altogether, it is about 1.5 meters long. Unlike the small intestines, the large intestines are not involved in the absorption of nutrients. Rather they absorb water and salts and pass the final waste material along to the anal canal, where that material is removed by defecation.

Digestion is a complex process with multiple steps. It takes some time. A lot of factors influence just how long food takes to pass all the way through the system, and there are differences among individuals and between men and women. Once food is swallowed, it takes about 6–8 hours to pass through the stomach and small intestine. The final steps of digestion, absorption of water, and elimination take about 36 hours in the colon. So in total, food might take 2 days or more to be fully processed and eliminated. Sometimes we might feel the need to visit the bathroom immediately after a meal. The gastrocolic reflex is a normal reaction to the arrival of food in the stomach. The final product is politely called feces, stool, or fecal matter. It is about 75% water and 25% solid materials. The solids include fiber, bacteria, cells, and mucus.

Foods eaten at the same time do not necessarily go through the system together. Some parts of a meal might be entering the colon while others are still in the stomach. Also foods different in how much time is required for processing. Juices or broth take about 15 minutes. Salad vegetables, such as lettuce and cucumbers, take about 30 minutes. Root vegetables, such as potatoes and carrots take about 50 minutes. Grains take about 90 minutes. Beans take a couple of hours. Meat and dairy products take the longest time. Some take up to 5 hours.

Our foods are digested down to their basic building blocks before they are absorbed into the blood stream. Protein digestion occurs in the stomach and the duodenum. Three enzymes are the primary actors: pepsin is secreted by the stomach, and trypsin and chymotrypsin are secreted by the pancreas. Their actions break the proteins down into smaller fragments called polypeptides. Various exopeptidases and dipeptidases then complete the breakdown to individual amino acids. Fat digestion begins with the action of lipase in the saliva, but its main digestion is completed in the small intestines by pancreatic lipase. Bile from the liver helps to emulsify fats so they can be absorbed as fatty acids. Triglycerides are broken down to free fatty acids and mono- and diglycerides, but no free glycerol is released. Starches or amylose are broken down by amylase from the saliva and pancreas. The result is simple sugars, such as glucose and maltose, that are absorbed by the small intestines. Milk contains lactose, and that is broken down by lactase. However, many humans lack this enzyme and cannot digest milk or milk products. Lactose intolerance is especially prevalent amount east Asian populations. Sucrase breaks down sucrose into its components of fructose and glucose, and those are absorbed by the small intestines. Nucleic acids are also degraded to mononucleotides by various nucleases that are secreted by the pancreas.

For some nutrients, special mechanisms are used to protect them and allow them to be absorbed intact. Vitamin B12 is a good example. In the saliva, B12 binds to haptocorrin, which protects it from digestion in the stomach. In the duodenum, the haptocorrin is removed, and the B12 binds to intrinsic factor. That complex is then absorbed in the ileum by a specific mechanism.

Those special mechanisms are rare. In the great majority of cases, nutrients are absorbed as their most basic elements. Supplements with collagen, a protein involved in connective tissue, are often recommended to improve joint function. These are typically available as a collagen hydrolysate in which the collagen has been digested by proteases to short peptides. Di- and tripeptides and even some larger peptides are absorbed and may find their way to joints, skin, and other sites of collagen in the body (Nogimura et al., 2020).

The undigested remnants of our food are eliminated by defecation. But the nutrients that are absorbed into our bloodstream are used by our cells for energy or repair parts. The waste products from the cells are ultimately transported back into our bloodstream and must be disposed of. The urinary system has that task. Embedded in that function are several other critical steps, including regulation of blood volume and blood pressure, maintaining levels of electrolytes and metabolites and maintaining blood pH (which is critical for gas exchange).

The kidneys are key organs in these functions. They are highly vascularized. Nephrons are the basic units of the kidney. An average adult kidney contains about a million of them. They filter the blood, reabsorb valuable materials (e.g., water, sodium, bicarbonate, glucose, and amino acids), and pass wastes (e.g., hydrogen, ammonium, potassium, and uric acid) into urine, which is drained to the ureters to the bladder and eventually voided by urination through the urethra. Humans produce 800–2,000 milliliters of urine each day, depending on how much fluid we take in and our kidney function.

When Things Go Wrong

Eating is second nature, and we rarely think about it—until something goes wrong. Our teeth hurt. We develop a canker sore in our mouth. We get heartburn or an ulcer. We suffer an attack of food poisoning or diarrhea. Finally, most of the tissues of the digestive system are subject to developing cancer.

Dental diseases are the most prevalent diseases (Heng, 2016). Although the fraction of people with cavities has decreased significantly in recent years, more than 90% of people aged 20 years and above have had a cavity. Tooth decay results from infection by *Streptococcus mutans*. Children get the bacterium from their mothers or other caregiver. The bacteria metabolize sugars to lactic acid, which then reacts with the calcium phosphate of the teeth. The weakened surface collapses inward to leave a cavity.

Bacteria growing on the teeth can form a biofilm (see Chapter 5) that facilitates the growth of *S. mutans*. Saliva flow is lower at night, and so, the antibacterial benefits of saliva are reduced, and the number of bacteria expands greatly overnight during sleeping (Sotozono et al., 2021). Bacteria are everywhere, and biofilms can form on essentially any solid surface (Saini et al., 2011). Teeth are a natural target. They provide a solid surface and a warm, wet environment with plenty of nutrients. More than 1,000 species of bacteria have been associated with dental biofilms. The bacteria form a solid three-dimensional extracellular matrix of their own secretions and other material from the environment. The biofilm needs to be reduced by ongoing dental care at home and regular visits to the dentist. If not, it can lead to tooth decay and gum disease.

Heartburn or gastroesophageal reflux disease (GERD) is one of the most prevalent disorders in the world, and it is increasing due to the aging of the population and the increase in obesity (Sweis and Fox, 2020). The reflux of the acidic stomach contents into the esophagus causes a sensation of burning behind the breastbone. Generally, over-the-counter treatments solve the problem. The two main ones are calcium carbonate tablets and proton-pump inhibitors (PPIs). American spend about $11 billion per year on PPIs. The cost of GERD in the United States alone was estimated to be $12 billion (Park, 2020).

Cancer is another disorder that can affect the gastrointestinal system. In fact, any of its parts and its ancillary organs (e.g., pancreas, gall bladder) can develop cancer, and those cancers account for more deaths than all other cancers combined. Those cancers show considerable affinities for specific areas. For example, esophageal and stomach cancer are much more prevalent in Asia and parts of sub-Saharan Africa. Pancreatic cancer is more common in the United States and Europe.

The prevalence of food allergies has also risen over the last decades (Blanchard, 2017). Food allergies involve an immune response to food and often involve immunoglobulin E and the release of histamine. They vary from mild (e.g., itching) to serious (e.g., vomiting, diarrhea, death). The most common foods associated with allergies are milk, peanuts, eggs, shellfish, fish, tree nuts, soy, wheat and sesame. The reason for the rise of food allergies is not clear. According to the hygiene hypothesis, early exposure to allergens might reduce the risk of developing food allergies (Ierodiakonou et al., 2016). Our lifestyles have changed radically over the last few generations and perhaps those changes have occurred too rapidly for our genetics to keep pace.

Probiotics are commonly advertised to improve immune responses, promote healthy gut flora, and treat and even prevent some illnesses. However, the stomach acid kills most bacteria, and a mixed reception in post-antibiotic recovery but only slowly and incompletely in the intestines (Han et al., 2021). The results of various studies have been inconclusive. For example, Suez et al. (2018) found that probiotics did colonize the gut of individuals in postantibiotic recovery, but they did it slowly and incompletely. However, autologous fecal microbiome transplantation resulted in a rapid and near-complete recolonization. While many commercial preparations include common bacterial strains, most studies have indicated that individuals require a more tailored approach if the treatment is to be effective (Veiga et al., 2020).

Diseases of Affluence

Our distant hunter-gatherer ancestors spent much of their time searching for food. Seasonal differences complicated the search. Once humans settled into villages and began agriculture, they still depended on enough rain and good weather to ensure a successful crop.

Today, in most of the developed world, we have controlled most of those issues. Our problem now is just the opposite: a world awash in calories. Food is plentiful. Supermarkets offer highly processed, prepared foods. Portions are super-sized, and everything comes with cheese. Sweetened drinks abound. It's far too easy to eat too many calories every day. Unfortunately, our bodies are still evolutionarily programmed to make the most of every calorie. A system that was beneficial for millennia now works against us.

In the future, diseases of affluence will be even more important (Choi et al., 2005). These result from urbanization, processed food, obesity, stress, pollution, and a sedentary lifestyle. Obesity and its complications, including diabetes, heart disease, and cancer, are already epidemic and will likely grow worse (Ward et al., 2019). Obesity also influences the immune system. The mechanisms are poorly understood. For example, Bapat et al. (2022) studied atopic dermatitis in lean and obese mice. They found that the obese mice progressed more rapidly to a much more severe disease state. In addition, treatments that were successful in the lean mice failed in the obese mice. Finally, too many people self-medicate with alcohol, tobacco, drugs, and food. Most tragic is that life expectancy has actually fallen already for some. For the last centuries, human lifespan has been increasing as we learned the importance of clean water and systems to deal with sewage and waste. We made great progress with antiseptics, vaccines, and other medical miracles. That great progress is now threatened by our own success.

9

Standing Up

That's one small step for man, one giant leap for mankind.

Neil Armstrong made his famous statement just as he stepped onto the Moon on July 20, 1969. There is some disagreement about what he really said. Was it ". . . for man" or ". . . for a man"? Armstrong believed he said the later, but most people remember it as the former. For him, it was just one step from the ladder of the space craft to the surface of the Moon, but it was a huge achievement for mankind to reach the Moon.

Most of us stand up many times every day. It's a trivial movement, but necessary for us to begin walking or for other purposes. However, it is as remarkable as humans walking on the Moon. From an evolutionary perspective, it involves the adoption of bipedal locomotion and the multitude of changes so that humans could walk efficiently on the ground. From a physiological perspective, it involves many tissues and organs, including bones, muscles, the heart, kidneys, inner ear, and much more. From our first wobbly baby steps, we use all of those systems to begin a lifetime of standing and walking.

Of course, humans are not the only organisms that stand up. Standing up is not absolutely required for every type of living thing, but in the broadest meaning, most organisms need some sort of support structures. Most animals need to move. Plants grow upward to gain more access to sunlight. Even individual cells need internal structures to move or to replicate. The structures take many forms. Some animals have internal skeletons like humans. Others have external skeletons or exoskeletons. Yet others eschew a hard support structure altogether. Plants rely on cellulose and other polymers to grow taller to efficiently collect sunlight for producing energy. Multiple support structures exist in other organisms and even in the cells that make up those organisms. This chapter will explore these and the "giant leap" that they all represent for humans and their species.

The Biology of Us. Gary C. Howard, Oxford University Press. © Oxford University Press 2024.
DOI: 10.1093/oso/9780197664797.003.0009

Standing on Two Feet

Bipedalism is one of the great advances in human evolution. Hominids began walking upright at least six million years ago, but even then, they retained many characteristics (e.g., long arms and short legs, long fingers and toes) that indicate that they spent a lot of time in the trees. The famous Australopithecine Lucy had a broad pelvis and thigh bones that put her feet in line with her center of gravity. By the time of *Homo erectus*, hominids were definitely bipedal. When humans walk or run, the feet take turns moving so that they are a half cycle out of phase with each other. In walking, the fraction of the time that each foot is on the ground is greater than 0.5. In running, it is less than 0.5. Humans are not the only animals that walk and run bipedally. Birds also walk and run on two legs, as do some lizards. The cockroach *Periplaneta* has several running strategies. It can run on six or four legs, but its fastest running is done on its two hind legs.

Why humans began to walk on two legs is not clear. Several hypotheses have been suggested. Monogamy might have encouraged males to walk upright so they could carry food to their family. Climate change might have forced early hominids to walk further to find food. Walking upright might have freed the hands to use tools or weapons. None of these arguments is compelling, and all are difficult to test. Other animals walk on two legs (Alexander, 2004), including apes, birds and many lizards. Some use forms of bipedal locomotion, including kangaroos, and rodents use a kink of skipping. Human walking is very efficient although running is less so, but the way other animals walk on two legs is significantly different from how humans do it. For example, chimpanzees bend their knees and their bodies. The differences add up to make walking on two legs energetically expensive for many of them.

Humans are the only bipedal great apes. We began as a quadruped, and many evolutionary steps were required to get us onto two legs. While the transition has worked well, it has not been perfect, and those changes also put us at risk of various musculoskeletal disorders. Bipedalism changed our knees and hips. As humans evolved, the knee structure changed to accommodate new stresses placed on it, and the genes responsible for those structures also changed. Bipedalism also changed our feet. Compared to our relatives, the great apes, we have a much larger heel bone, more-developed arches, and shorter toes, and unlike our thumbs, our big toe or hallux is not opposable. Most of the apes can climb far better than we can, but all these changes allow us to walk more efficiently on two feet. We use our feet as stiff levers, whereas apes have a flexible midfoot joint. Traditional discussions of the evolution of human feet in bipedalism have focused on the dichotomy of climbing apes and walking humans. In an excellent review, Holowka and Lieberman (2018) take a broader, more nuanced view that includes more possibilities for adaptations, such as running and fighting. They look at foot strike, push off, and bioenergetics and found some similarities in human and ape feet during walking. They divide the evolution of human feet into three stages: a foot suited for climbing, one that could both climb and walk, and finally one adapted to walking and running. The ankle also has a role in bipedal evolution. The muscles and tendons of the ankle produce power to assist in walking called ankle push-off. That power might be used in swinging the leg or accelerating the center of mass. Zelik and Adamczyk (2016) suggest that it is used in both. They suggest that push-off increases the speed of the training limb that provides the push-off. However, it also accelerates the center of mass.

Whatever the advantages of bipedalism for humans, there was clearly a cost, and that is still being felt today. The muscle-skeletal rearrangements continue to be felt in the increasing incidence of osteoarthritis (Richard et al., 2020). Selection pressures and more recent actions have combined to increase the risk of osteoarthritis. For example, chondrocytes are the cells that produce the collagen and extracellular matrix that are key to the cartilage that protects the knee joint. The epigenetics of noncoding DNA sequences for chondrocytes may contribute to modern knee problems (Richard et al., 2020). The human skeleton has been studied carefully by paleoanthropologists, but Kun et al. (2023) contributed a new approach. They used deep-learning models to examine more than 30,000 full-body x-ray images to determine the genetic variations in human bones. Their findings were interesting. The proportions of limbs did not correlate with body width, but those of hips and legs did associate with osteoarthritis. Finally, genes for skeletal proportions were associated with regions related to human evolution. An estimated 240 million individuals worldwide suffer from this condition, including 10% of men and 18% of women 60 years and older (Allen et al., 2022). Osteoarthritis in the knee is a common disorder that affects 14–30 million people in the United States. Its symptoms include inflammation, pain, and functional limitations. It is managed by nonsurgical treatments, such as exercise and weight loss, pain relievers, or injections of corticosteroids and hyaluronic acid (Tolpadi et al., 2020). The only treatment for more severe disease is a total knee replacement, and about 400,000 of these procedures are done each year. The outcome is generally good, but some fraction of patients complain of problems.

Cells and Cellular Structures

The need for support structures begins at the cellular level. All living organisms either exist as a single cell or are made up of many individual cells organized into organs and tissues, and the most fundamental support structures are inside those of individual cells. We humans have a very large number of cells. Estimates of that number vary and are complicated by the myriad "fellow travelers" that we carry. One estimate is 3×10^{13} human cells (Sender et al., 2016). Individual cells do not stand up, but some can move, and they often require specific structures for their functions, such as transport of materials within the cell, motility, transmitting force, reacting to external forces, changing the cell's shape, and cell division (Fletcher and Mullins, 2010; Hohmann and Dehghani, 2019). Even amoeba, which are famously known to be blobs of protoplasm, have some structure. These deceptively simple cells move across the surface by extending and retracting projections from the main body called pseudopods. Also, cells divide for at least part of their life, and keeping that organized requires structure and contractile mechanisms. The human genome has 3 billion base pairs arranged onto 23 pairs of chromosomes (Figure 9.1). That amounts to 2 meters of DNA in every cell. Mitosis is a highly structured event indeed. Each cell has a structure called the cytoskeleton. The cytoskeleton is a combination of various protein-based elements. The three primary types are actin, microtubules, and intermediate filaments. Those filaments give the cell its shape. It helps the cell to migrate to new positions. It helps the cell membrane to engulf extracellular material, and allows chromosomes to be segregated during mitosis, and aids cell division.

Slime molds are fascinating creatures. Their name is partially correct. In one phase of their lives, they do look slimy, but they are not fungi. They are now in the group Protista. During a long career at Princeton, John Tyler Bonner studied slime molds. He described them as "no more than a bag of amoebae encased in a thin slime sheath. Yet they manage to have various behaviors that are equal to those of animals who possess muscles and nerves with ganglia, that is, simple brains" (MacPherson, 2010). In recent

Figure 9.1. Separation of chromosomes during cell division. The chromosomes (blurry masses marked by dots in center) are drawn apart by the spindle apparatus to the kinetochores at the opposite poles. The spindle apparatus is composed of contractile proteins.

Reproduced from Afunguy. (2006). Kinetochore. https://commons.wikimedia.org/wiki/File:Kinetochore.jpg. Public domain license.

years, slime molds have been studied for their ability to solve mazes and networks (Patino-Ramirez et al., 2021). However, they have a unique manner of standing up. When food is plentiful, *Physarum polycephalum* live as individual single-celled organisms. But when food is scarce, the individual cells aggregate into a syncytium, a single multinucleate cell, that begins to move in search of food. Also, they are not small. *P. polycephalum* can be a meter or two in diameter. Under certain circumstances, they form a fruiting body. The fruiting body consists of a stalk of dead cells, capped with a bulb of cells that produce spores. Those spores are dispersed by wind to form new individual cells at some distance from the original organism. If the new organisms land in a moist environment, they develop into amoeba, but if they land in an aqueous environment, they develop into flagellated cells. The movement of slime molds is also interesting. Individual cells move in an amoeboid manner by extending and retracting pseudopodia. When they form a single organism, they also use pseudopodia to explore their environment. They are attracted to food (e.g., sucrose) and withdraw from other substances (e.g., salt). Slime molds are not slow and can cover a few meters in 24 hours. Amazingly, they can "remember" where they have been.

Slime molds use several strategies that allow them to transform from single cells to multicellular structures. *Dictyostelium discoideum* live as single-cell amoebae, but under some circumstances, the amoebae begin to secrete cyclic AMP, which forms a gradient of the molecules. Chemotaxis causes the cells to move toward the cell secreting the most cyclic AMP and aggregate. The cells then differentiate into either stalk or spore-producing cells. Apoptosis, or programmed cell death, causes the cells that form the stalk to die. *Hemitrichia serpula* forms a multicellular blob called a plasmodium. The cells undergo mitosis, but without cytokinesis. In other words, the nuclei divide, but the cells do not separate. The plasmodium moves around in search of food and can also develop into fruiting bodies that release spores. The cellular slime mold *Fonticula alba* lives as an amoeboid single-cell organism. Under certain circumstances, the cells aggregate to form a mound. The cells then secrete an extracellular matrix that changes the mound into a tapered stalk that looks like a volcano. At the top of the stalk, cells form spores that are later released (Brown et al., 2009; Worley et al., 1979).

Animals

Animals use a variety of strategies for support. Vertebrates, such as humans, have an internal skeleton, but invertebrates get along just fine with an external skeleton or exoskeleton or no skeleton at all. Biology is amazingly inventive. If you can imagine a creature, chances are that nature has already tried that combination.

Slugs and worms have no rigid tissues and move by using their muscles to push against a substrate. Typically, caterpillars move by inching or crawling (Griethuijsen and Trimmer, 2014). Inching is not well understood, but crawling is. Caterpillars have a couple of interesting mechanisms. First, they grip the substrate strongly and, essentially, use it as a "skeleton." Second, the caterpillar's body is quite elastic, and they use their substantial gut to assist in movement. The gut and body wall move independently to assist crawling. Earthworms also lack any type of skeleton. *Lumbricus terrestris* is a common example. Their basic structure is that of a tube within a tube. The outer tube is the outer layer of the animal, and the inner one is the gut, which runs the length of the worm. Earthworms are segmented worms, and that means that each segment is the same as the others except for a few with specific modifications. Each segment has circumferential and longitudinal muscles, and these are used to move the worm as it wriggles along. Another set of these muscles lines the gut and moves food along as it's digesting.

Octopuses are extraordinary animals. They belong to the phylum Mollusca and the class Cephalopoda and so are closely related to squids and cuttlefish. They have two excellent eyes, eight legs, and a rostrum or "beak" that they use like jaws. Most of the Mollusca have some sort of hard shell (e.g., snails, clams, oysters). However, the only hard part in the octopus is the beak. This small radula is a ribbon-like structure with tiny teeth and made of chitin, similar to that used by snails. John Steinbeck (1945) described octopuses as, "the creeping murderer, the octopus . . . pretending now to be a bit of weed, now a rock . . . runs lightly on the tips of its arms, as ferociously as a charging cat." Octopuses move in different ways. They typically "crawl" along the sea bottom by extending their legs and pulling themselves along. However, they can also move quickly by expelling water through their siphon in a sort of jet action. Interestingly, they are the only animals without bones that can "walk" on two legs, using a hydrostatic "skeleton" rather than the rigid support of bones or an exoskeleton. To walk, the octopus uses the transverse, longitudinal, and oblique muscles in each arm. Those muscles are flexed or relaxed, but the internal volume of the arm remains constant. A bend in the arm is propagated down the arm to propel the animal forward (Huffard et al., 2005; Huffard, 2006). This behavior has been documented in at least three species (*Amphioctopus marginatus*, *Abdopus aculeatus*, and *Octopus vulgaris*) (Huffard et al., 2005, Amodio et al., 2021). Scientists speculate that this walking strategy is used by the octopuses to escape predators. Sharks and other predators recognize the typical crawling behavior and target the octopus. By walking on two legs, the octopus can disguise itself as a clump of algae or a broken coconut shell and fool the predator.

Exoskeletons come in two general types. Snails, clams, oysters, and mussels have very heavy exoskeletons of calcium carbonate called shells. Crustaceans and insects have a softer exoskeleton made of chitin, with occasionally some amount of calcium carbonate. Regardless of their composition, the exoskeleton provides protection from predators and desiccation, structure for muscle attachments, and feeding. Exoskeletons include many different minerals, including silica, apatite, and carbonates. Aragonite and calcite are two crystal forms of calcium carbonate. Each group uses its own mineral combination, and that material seems not to vary. However, some scientists hypothesize that the specific composition reflects the seawater that they were in when they first evolved. The hypothesis is tricky to test because, for example, aragonite recrystallizes to calcite at normal surface atmospheric pressures, so a calcite skeleton might have originally been aragonite. Aragonite fossils older than the Carboniferous are exceedingly rare. Aragonite deposits are typically found on the ocean floor. Nevertheless, the correlation of skeletal types correlates with the local seawater chemistry when that group first evolved (Porter, 2007) and lends support to the basic hypothesis.

Exoskeletons have advantages and disadvantages. Their rigidity is a significant challenge to growth of the organisms, and some organisms, such as arthropods (e.g., insects, spiders, crayfish) shed their shell to grow. As one example, each summer, cicadas (*Neotibicen linnei*) emerge to call to one another at dusk and dawn. They leave their discarded skeletons in trees and other places. Also, crabs periodically shed their exoskeleton. While their new shell is hardening, they are vulnerable to predation, and they typically stay hidden until the new shell is ready. We humans call them soft-shell crabs and enjoy them for dinner. The exoskeletons of arthropods contain chitin and proteins (Zhao et al., 2019). This very strong material is a polymer of a derivative of glucose. Its chemical name is N-acetylglucosamine. Chitin is common in nature. In addition to arthropods, it is found in the cell walls of fungi, the radulas of mollusks, the beaks of octopuses, and in some nematodes and diatoms. The chitin is sometimes supplemented with calcium carbonate for added strength.

Other organisms (e.g., clams, oysters, mussels) have open exoskeletons so that growth can continue by simply adding new material to the edge of the shell. However, those shells are very heavy and limit the mobility of the organism. They are made of calcium carbonate and other salts as well as proteins and polysaccharides. Diatoms and radiolaria use silica to form their exoskeletons, which are beautiful crystal structures when viewed under the microscope. Foraminifera are small single-celled protists. Most live deep in the ocean, and they are quite beautiful to look at under a microscope. In a quite different strategy, their shells are not made from materials produced by the organism, but from materials collected from the local environment and agglutinated onto the organism. They have shells. The shells vary, with one or more chambers that have holes. The shell might consist of calcite, but the most fascinating are made of sand grains or other material that are agglutinated into an organic layer. For example, *Spiculosiphon* prefers to embed silica spicules from sponges in an organic cement (Maldonado et al., 2013). Xenophyophores are the largest foraminifera with diameters up to 20 cm. They add small grains of sediments, such as sulfides, oxides, volcanic glass, and even the shells of smaller foraminifera.

Interestingly, turtles, which are vertebrates and reptiles, evolved to use both an internal and external skeleton (Cordero, 2017). Their back shell is bone that grows from the ribs to form flat plates and is connected to the vertebrae. The front or belly plate grows from the should girdle, front ribs, and sternum. The bony plates are covered by epidermal scales. The turtle's shell provides shelter and protects the animal. Unlike insects, turtles do not shed their shells. As the turtle grows, so does the shell. Turtles are not the only vertebrates to use a form of exoskeleton. Pangolins and armadillos have hard outer coverings. In armadillos, the armor is plates of bone covered by epidermal scales of keratin. In pangolins, the plates are made of keratin. Several groups of dinosaurs also featured bony armor. Much of the body of Ankylosaurus was covered with plates. The tail of Stegosaurus had bony plates and ended in spikes. Triceratops had bony armor on its head, a bony frill, and three horns.

Humans

One of the thrills of being a parent is to watch a baby take its first steps. The event is preceded by several preparatory stages, such as belly crawling, hands and knees crawling, and "cruising" (moving while holding on to something) (Adolph et al., 2011). Each baby develops in its own way, and through all of this work, babies are strengthening muscles and gaining control of their balance. By 10–18 months, most babies have put it all together and started on the long journey of their life (Størvold et al., 2013).

Humans and other vertebrates depend on an internal skeletal system to provide a basic framework to support the animal and to protect vital parts. Adult humans have 206 bones. Babies have a few more, but those fuse together later in development. The initial separations allow for growth. Males and females differ somewhat in terms of the bones. The female skeleton is smaller than the male, and the female pelvis is wider to accommodate childbirth. The bones themselves are 40% inorganic material (hydroxyapatite or calcium phosphate), 25% water, and 35% organic material (mostly collagen type 1 and other proteins) (Feng, 2009).

In addition to the bones, the skeletal system comprises the muscles, cartilage, tendons, ligaments, joints, and connective tissues of the body. The bones provide a structure that enables the body to stand up and move and also attachment points for the muscles to pull against. Tendons and ligaments are examples of strong connective tissue. Tendons connects muscle to the bone, and ligaments connect bone to bone. The bone-on-bone connections are padded by cartilage, a flexible connective tissue made primarily of collagen. As noted in an earlier chapter, muscles contract and relax. As a quick review, muscles are made up of several proteins. The main players, actin and myosin, each form long fibers that are interlaced. When the muscle is activated, the myosin fiber binds to the actin and contracts. Then it releases the actin, takes a new grip, and contracts again. Later it relaxes. The key here is that muscles only pull. They do not push.

Of course, it takes more than bones and muscles to stand up. The heart is among the very first organs to develop in embryos. The other organs and tissues need the nutrients and gas exchange that are facilitated by the circulation. The pumping of the heart moves blood throughout the body, and that action generates a blood pressure that varies with the demands of the individual organism. The pressure has several components, including the diameters of the blood vessels (especially the small ones) and the stiffness of the arteries. Schulte et al. (2015) reviewed the evolution of blood pressure from when our ancestor lived in the sea, then came on land, and further evolved systems that required high blood pressures and yielded great endurance. Simple physics has a role. When we are sitting or standing, the heart has to create sufficient pressure in the arteries to move blood "uphill" from the heart to the brain. For us, that hydrostatic head is typically about 40–80 centimeters. But for some animals, those numbers can be much greater. Giraffes have very long necks and thus require a much higher blood pressure. Arboreal snakes have higher pressures than terrestrial snakes. The long-necked dinosaurs, such as Apatosaurus, Brontosaurus, and Barosaurus, would have required very high blood pressures (Black, 2013; Seymour, 2016). Supersaurus, the largest dinosaur so far, had a neck 50 feet long. Calculations of the pressure required and the anatomy of the giant indicate that they normally held their heads close to the ground. They seemingly did not have the ability to raise their heads to the full length. One complication of the blood pressure required to stand up seems to involve fainting. That type of fainting is called orthostatic fainting (van Kijk, 2003). Interestingly, other animals (e.g., giraffes and arboreal snakes) do not faint. Speculation about this tendency in humans centers around the larger amounts of blood needed for the human brain and our large legs that pool a lot of blood.

Liu et al. (2021) compared the genome of giraffes to other ruminants and found a series of mutations that were specific to giraffes. Those mutations were related to cardiovascular and other functions. A mutation in the gene FGFRL1 has seven unique amino acid substitutions. When those substitutions were made in a mouse model, the mice had great resistance to hypertension and higher bone mineral densities. Those traits would be of great value to the giraffes.

Regulation of blood pressure is complicated. While it is not completely understood, several components are known. Baroreceptors detect changes in arterial pressure and signal the medulla portion of the brain. The proteins renin and angiotensin cause the arteries to contract or relax to help regulate pressure. The hormone aldosterone controls stimulates the kidneys to retain sodium and excrete potassium, and salt and water levels control plasma volume and, indirectly, blood pressure.

Standing and moving are conscious activities. In most cases, we think about standing up, and so, that action requires the functions of our nervous system, including the brain and the nerves that connect our brain to our muscles. The mysterious relationship of consciousness to the physical brain is the subject of another chapter. The physical and biochemical aspects of the nervous system are far better understood. The key players are cells called neurons, and there are a lot of them in the brain and in the peripheral nervous system. They "connect" to each other at synapses. In fact, they don't really connect. There is a tiny gap between them, and they communicate by the diffusion of neurotransmitters that are released from one side and picked up on the other. Thus, a stimulus (e.g., a thought) causes a change in sodium and potassium ion concentrations inside and outside of a long projection of the neuron called the axon. The rapid concentration changes allow an electrical signal to be promulgated along the length of the axon to the synapse where it causes the release of the neurotransmitter. That chemical diffuses across the synapse and binds to the next neuron, which then activates its own electrical transmission to the next neuron or ultimately to the muscle. All of these allow us to stand up, but it is just as important to remain standing, and that requires mechanisms to maintain our balance. On February 26, 1964, former astronaut and aspiring US senator John Glenn slipped and fell in his bathroom. He suffered a concussion and damage to his inner ear that left him with severe vertigo. Our sense of balance depends on the coordinated work of the eyes and the inner ear to help us determine our position in a three-dimensional world. The inner ear contains five sensory organs that connect to the vestibular nerve (Spoor et al., 2007). Put simply, the movement of fluid in the inner ear along with a small bulb stimulates small hairs to signal the brain. Here is a somewhat more technical explanation. Three semicircular canals are thin tubes that contain a small gelatin bulb called the ampullary cupula. It is connected to hair cells and bathed in endolymph. Movement of the cupula in the endolymph causes the hair cells to send messages to the vestibular nerve about the movement. In addition, two otolithic organs help to sense movement by stimulating cilia that transmit signals to the brain. Damage to these sensitive organs causes a loss of balance and vertigo, such as that suffered by John Glenn. Those small hairs in the inner ear also help us to tell up from down.

The bones, muscles, blood pressure, nerve impulses, inner ear, and the rest are only some of the many physiological systems that must work in concert to enable us to stand up and move. Yet we routinely manage all of these various systems without a thought.

Plants

Plants in the water are buoyed up by the water. As a result, the walls of the plant cells are thinner than the rigid walls of land plants, which have to support the weight of the plant and also withstand harsh weather. Aquatic plants that grow in water can be loosely divided into three groups, according to how they support themselves. Plants generate energy by photosynthesis, and so, they need good exposure to sunlight. Water absorbs sunlight. The deeper in the water, the less sunlight is available. Free-floating plants live on the surface of the water and are not rooted in the bottom of the body of water. Emergent plants are rooted in substrate but grow above the water surface. Some use their buoyancy to float on the surface and then grow taller. Others have support structures that allow them to grow above the surface. These are often found in wetlands and along the shore where the water is 1–2 meters deep. The final group is submerged plants. These are rooted in the substrate but depend on their buoyancy to reach the surface. Most are found in water that is 3–6 meters deep, but some are much larger. The submerged group includes several species of kelp (e.g., *Alaria esculenta*, *Laminaria digitata*, *Saccharina longicruris*, *S. latissimi* on the Atlantic coast and *Nereocystis luetkeana* or bullwhip kelp on the Pacific coast). Kelp are brown algae that form large colonies that resemble underwater forests. Bullwhip kelp grows to 36 meters in length. At the bottom is a thallus or holdfast that secures it to the substrate. At the top is a pneumatocyst that sprouts 30–60 blades that are up to 4 meters in length. The pneumatocyst is filled with carbon monoxide and buoys the plant up so that the blades can absorb more sunlight in the upper waters (Liggan and Martone, 2018).

Aquatic plants have characteristics that allow them to survive and thrive in the water. They typically feature flat leaves and hollow roots. Most importantly, they have air sacs that allow them to float in the water (Sloane, 2019). The leaves of the common bladderwort (*Utricularia macrorhiza*) contain air spaces in the shoots. The water primrose (*Ludwigia* sp.) has two types of roots. One attached the plant to the substrate. The other features air sacs and are adventitious.

To move from the oceans to the land, ancient plants had to evolve strategies that allowed them to overcome numerous problems. Standing up to gain exposure to the sun required new support systems that did not depend on the buoyancy in water. They needed to protect themselves from drying out in the air. They also needed to develop new ways to reproduce in a dry environment. Of course, none of these new systems had to be developed overnight. Plants likely grew into area of more and more shallow water over considerable time. Here we will focus on support. The first plant "explorers" of the land are thought to have evolved from the streptophyte green algae. Those algae developed several key characteristics required for the movement onto land (Becker and Marin, 2009). For example, the phragmoplast is a complex assembly of microtubules, actin filaments and associated molecules. This plant cell specific structure appears in late cell division and provides a framework so the cell wall can form to separate the two daughter cells. They also contain hexameric cellulose synthase, which is needed to synthesize cellulose to form the strong cell walls.

Among all plants, only land plants have been able to stand up above the substrate (de Vries and Archibald, 2018). Land plants can be a centimeter tall or as much as 100 meters. On the small end are the liverworts. They typically grow as a flattened moss or flat thallus without roots, stems, a vascular system, flowers, seeds, or leaves. They have single-celled rhizoids instead of roots. About 9,000 species of liverworts are known, but they are small and easy to overlook. Liverworts are widely distributed but prefer moist environments. Because they are so small, they do not have any significant support structures. Hornworts and mosses also grow very close to the ground. At the other end of the spectrum are the redwoods (*Sequoia sempervirens*). They are the tallest trees on Earth and reach 115 meters in height and nearly 9 meters in diameter.

Trees, as a group, are the largest of plants. Their support structure is provided by the woody material under the bark. Most of this material is actually dead (Parham and Gray, 1984). It is the remnants of the xylem and is composed of multiple polymers. The major component is cellulose, which is a polymer of sucrose. The cellulose is embedded in a matrix of lower-molecular-mass polysaccharides that include different five-carbon sugars (e.g., xylose, arabinose) and six-carbon sugars (e.g., glucose, mannose, galactose). They are referred to as hemicelluloses, and they are all hydrophylic. The third component is lignin. It consists of highly branched polyphenolic molecules. It is rigid and gives strength to the wood. Lignin is hydrophobic. The sail effect on trees in heavy winds can result in considerable stress on the tree. Trees of different species react in different ways. In urban and suburban areas, the failure of trees can cause great expense and even injury if the mechanical stability of the tree is exceeded. Thus, efforts have been made to better understand the processes beyond that of the eye of an arborist. Size, canopy shape, and branch structure complicate the determination of those stresses. In particular, the branches move somewhat independently. In general, trees begin to fail at windspeeds above 60 miles per hour. At 150 miles per hour (a strong hurricane), the leaves are stripped. James et al. (2006) developed a dynamic model to predict tree survival.

Cellulose and lignin allow trees and other plants to grow tall, but they must be firmly anchored into the ground. Roots also absorb water and nutrients from the soil so they can be distributed throughout the plant (McCully, 1999). Root growth is governed by the environment. Most roots occur relatively near the surface where aeration and nutrients are most available. Roots also take many forms and specializations. Some are storage organs (e.g., carrots, beets, sweet potatoes). Some allow the plant to propagate vegetatively. Others grow along the ground to buttress the plant. Root nodules are enlargements on the roots that contain nitrogen-fixing bacteria. These are extremely important. Although nitrogen makes up the majority of the air we breathe, it is not available chemically, and all living organisms need nitrogen for nucleic acids, amino acids, and other critical biomolecules. Thus, the nitrogen-fixing bacteria are crucial for life on Earth.

All plants need water. They get some by absorption through the leaves, but most must be transported from the ground (Merhaut, 1999). In redwoods, for example, water has to be lifted a very great distance. How do they do that? Plants have an extensive plumbing system, consisting of channels called phloem and xylem. As those channels extend up through the plant, they diverge into smaller tubes that reach every part of the plant. As was noted in Chapter 6, water is a polar molecule. That means that water molecules have negative and positive ends, somewhat like a miniature magnet. Even that tiny charge causes water molecules to bind to each other, and this property explains some unusual behaviors of water. Water beads up on polished surfaces. It seems to defy gravity and to climb up a paper towel. This phenomenon is called capillary action. Plants take advantage of capillary action to draw water up from the ground to the upper branches and leaves. The water flows upward through small tubes called the xylem.

Vines are plants that outsource their support systems to another plant or object. They cling to the other structure and take advantage of that support to grow upward. Some plants (e.g., poison ivy or *Toxicodendron radicans*) grow as a small shrub but will transform into a vine if external support becomes available. Others (e.g., honeysuckle or *Lonicera japonica*) are always a vine.

Vines have various strategies for climbing their support. Some wrap themselves around the support. Some have adventitious roots that grab the support. Some have tendrils that are like curly threads or wires that hold onto the support. The tendrils might also have adhesive pads that stick very firmly to the support. Others use hooked structures to attach to the support. Some vines are highly desirable. They can be used to good advantage as a decorative element. Some people like to grow climbing green beans among their corn so the beans can use the corn stalks as support. Grapes and roses are also good examples. Others vines are pests. Poison ivy causes severe allergic reactions in most people, and kudzu (*Pueraria* genus) has invaded the southeastern United States with devastating consequences for the forests there.

How do plants determine up from down and remain standing straight? They are subject to a multitude of forces, including gravity, wind, rain, and animals passing by. Even their growth is nonlinear with many curls and twists. The answer is a small starch-filled grain called a statolith in cells called statocytes (Berut et al., 2018). These grains collect at the bottom of a cell and respond to even the slightest change of angle. Upon sensing a change in inclination, these sensors activate subtle changes in growth hormones. However, they do not sense changes in gravity, but rather sense changes in inclination. Moulia et al. (2021) term the overall sensing of position by plants as proprioception. Plants use many cues to sense and adjust their shape, growth, and development.

Moving Plants

Macbeth, in the eponymous play by Shakespeare (Act 4, Scene 1), sought help from three witches, who conjured up three apparitions to tell his future. He was greatly comforted by the prediction of the third apparition—that he would be safe until the forests of Great Birnam Wood marched against him. "Macbeth shall never vanquished be, until Great Birnam wood to high Dunsinane hill Shall come against him." How would trees move from one place to another? Animals move and plants do not. That simple rule made sense to Macbeth and to most of us as well.

However, the rule is not perfect. Some animals move so little that they are often mistaken for plants. After a motile early phase, most sponges and corals settle down on a surface and remain there. Plants are not usually associated with movement, but biology being biology, it does occur. Some plants move at least a little sometimes.

The science of plant movement began with Charles Darwin. His book *The Power of Movement in Plants* (Darwin, 1880), summarized his experiments studying this phenomenon. Most of us are familiar with some types of plant movement. Plants grow toward light, and flowers "close" their petals at night. Some plants move their leaves when touched, and some carnivorous plants have mechanisms that snap traps to capture insects (Bauer et al., 2021).

Indoor plants often must be turned so that they grow symmetrically. Plants grow toward light to maximize energy production under changing environmental conditions (Liscum et al., 2014). This phototropism involves the cell elongation that directs the plant to the light. Six photoreceptors are involved along with their respective signaling pathways. The initial signaling results from light that is detected at the cell membrane and followed by additional signals in the cytoplasm and nucleus. Exposure of photoreceptors located in the shoot tip above the growth zone to blue light causes the movement of the hormone auxin. Auxin migrates from cells exposed to the light to those less exposed and also downward toward the roots. As a result of the excess auxin, cells on the shaded side elongate more than those on the lit side and cause the plant to bend toward the light.

The sensitive plant *Mimosa pudica* reacts to a touch. The leaves fold inward and bend down and then open after a few minutes. These thigmonastic movements are controlled by electrical and chemical signals that spread the alarm throughout the plant. A stimulus (e.g., touch, warming, shaking due to the wind) causes mechanoreceptors receptors to initiate an action potential involving changes in calcium and other ion concentrations, similar to that in animal cells. The action potential progresses from the leaflet to the stem and on to other leaflets. The key structure in the plant movement is the pulvinus. This swollen area occurs at the end of the petiole where the leaf connects to the stem. In the pulvinus, ion changes cause the loss of water (or loss of turgor) in the leaves that causes the leaves to move (Song et al., 2014). Some carnivorous plants have similar mechanisms. For example, the Venus flytrap (*Dionaea muscipula*) has trigger hairs that cause the trap to snap shut when a victim lands on the flower (Hedrich and Neher, 2018).

Flowers "go to sleep" by closing their petals at night. They use pulvini to move the petals, but the mechanism that initiates the process is a bit different from the one that underlies sensitivity to touch. Sleep or nyctinastic movements are controlled by the plants' circadian rhythm and the ratio of red and far-red light in their environment. Leaf-opening or leaf-closing factors are also involved (Ueda and Nakamura, 2007). Their binding sites are on the surface of the motor cell, which shrinks or expands in response to the factor. The opening and closing of flowers is related to fertilization and reproduction (van Doorn and Meeteran, 2003). Flowers open when pollinators are active and close when they are absent. The opening is surprisingly quick: it varies from 5 to 20 minutes. In most species, flower petals open and close with differences in growth of the two sides of the petals. To open in the morning, the inner surface grows rapidly as the temperature goes up, but the outer surface does not. As the temperature cools, the outer surface speeds up. The difference in the optimal growth temperatures of the surfaces is about 10°C. In other plants, the movement is in response to changes in turgor pressure along the midrib rather than the petal surfaces.

Chinese witch-hazel (*Hamamelis mollis*) shoots its seed for several meters (Poppinga et al., 2019). The outer layer of the fruit shrinks, and the inner layer both shrinks and expands. That combination puts enormous force on the seeds that finally releases the seed with the force of a bullet. The action may also involve the movement of sucrose or potassium ions between cells that changes water flow and turgor pressure that add to the energy that expels the seeds.

Putting It All Together

Standing and walking are so much a part of our everyday lives that it is easy to forget just how profound those movements are. A baby's first steps recapitulate the first steps that hominids took millions of years ago on the plains of Africa. The bones, muscles, blood pressure, nerve impulses, inner ear, and the rest are only some of the many physiological systems that must work in concert to enable us to stand up and move. Yet we routinely manage all of these various systems without a thought. Of course, with so many systems involved, something is likely to go wrong. Even a slip in the bathroom can result in disaster, as John Glenn found. Broken bones, high or low blood pressure, heart disease, osteoporosis, loss of muscle mass with aging, neurodegenerative diseases, and more can make standing and walking a challenge or even an impossibility.

Why early humans began to walk on two feet is not known, but those steps deeply influenced the development of humans. Bipedal locomotion might have allowed humans to more efficiently gather food for improved nutrition and further brain development. It might have freed up our hands to use with tools or weapons. As those many early human babies continued to walk, they carried us out of Africa to populate the world and then to allow Neil Armstrong and others to walk on the Moon.

10

The Heart of the Matter

For it is by the heart's vigorous beat that the blood is moved, perfected, activated, and protected from injury and coagulation. The heart is the tutelary deity of the body, the basis of life, the source of all things, carrying out its function of nourishing, warming, and activating body as a whole.

—William Harvey (1628, p. 59)

A slip with a kitchen knife. A nick while shaving. A cat scratch. And suddenly, we are bleeding. Who hasn't suffered one or another of these minor accidents or even something worse? Almost miraculously, in a few minutes in most cases, the bleeding stops. Blood is amazing, but clotting is only one of it critical functions. It carries oxygen and nutrients to all our cells, and it carries aways carbon dioxide and waste products. It transports signaling molecules, hormones, and other important molecules.

As William Harvey noted, it is the heart that moves the blood. "The heart is the tutelary deity of the body." Blood and the heart have special significance to humans. It is used in many words. The heart was long considered the seat of emotions and even the soul. Even today we speak of heartache or broken hearts or winning someone's heart. Hearts are a common theme on Valentine's Day. Blood has also had emotional connotations, including blood relative, blood lines, and blood brothers. Heartbeats and blood are traditionally ways to determine if someone is even alive.

However, here we will stick with the biology and leave the philosophy, emotions, and images to others. We will review the composition, evolution, and diseases of the blood and how it is moved, oxygenated, and purified. Thus, we will cover the organs that are closely involved with blood. Our four-chambered heart moves blood throughout the body. The lungs ensure that the blood is efficiently oxygenated, and the kidneys clean impurities from

The Biology of Us. Gary C. Howard, Oxford University Press. © Oxford University Press 2024.
DOI: 10.1093/oso/9780197664797.003.0010

the blood and help to manage blood pressure. In this chapter, we will look at each of these, beginning with blood.

Blood

Blood itself is a miracle fluid. The average human adult has about 5 liters of blood. As a rule, women have a little less than men. This complex mixture includes plasma (54.3%), red blood cells or erythrocytes (45%), white cells of various types (0.7%), and many proteins and other molecules, such as antibodies, lipoproteins, hormones, and other signaling molecules.

Composition

Once the cells have been removed from the blood, the remaining fluid is plasma. This straw-yellow-colored liquid comprises mostly water (92%) with many important proteins, glucose, amino acids, fatty acids, and salts (e.g., sodium, chloride). It also contains waste products, such as urea, lactic acid, and carbon dioxide. Another fluid phase of blood is called serum, but serum and plasma are actually quite different. Serum is plasma that has had the clotting factors removed. In practice, this is accomplished by allowing the serum to clot and then subjecting the mixture to centrifugation to remove the clotted material. Plasma contains many important proteins. Serum albumin is a carrier protein for steroids, fatty acids, and thyroid hormones. It's also critical for maintaining the osmotic pressure of the plasma and for binding calcium. Plasma also contains blood-clotting factors, immunoglobulins, and lipoproteins that will be further described below.

Platelets are the first line of defense against bleeding. These small cell fragments lack a nucleus. They circulate freely until they detect bleeding. They plug small wounds, but aggregate and react with fibrin to form a clot for larger injuries. There are a lot of them. The average adult produces about 10 billion platelets every day. They last about 8–9 days in circulation, and then they are destroyed in the liver and spleen. Platelets come in a variety of forms and functions (van der Meijden and Heemskerk, 2019). They differ in size, content, glycoprotein expression, age, and receptors. Larger platelets contain membrane structures, such as endoplasmic reticulum, mitochondria, lysosomes, and secretory granules. Finally, platelets even from a single individual can differ greatly in function.

Red blood cells (erythrocytes) contain hemoglobin and carry oxygen. They are a major component of the blood and make up 45% of blood volume. A drop of blood contains about 5 million red blood cells. Under the microscope, they are biconcave discs. That is, they look like little doughnuts without a hole. They are red because of the hemoglobin that they contain. Each cell contains about 270 million molecules of hemoglobin. They are also unusual among cells in that they lack a nucleus and other organelles. Red blood cells develop from hematopoietic stem cells in the marrow of large flat bones mostly (e.g., pelvis, vertebrae, ribs, breastbone) (Moras et al., 2017). Like many cells that are in constant usage, they have a limited lifespan of about 100 to 120 days in circulation. Once they wear out, they are recycled, also in the bone marrow. To maintain an adequate number of red blood cells in the circulation, about 2 million are produced every second. Interestingly, the kidneys monitor the number of cells in circulation. If there are too few, the kidneys make and secrete the hormone erythropoietin that stimulates the marrow to make more cells. The cells pass through several stages to become red blood cells. The hematopoietic stem cell becomes a common myeloid progenitor cell that has the ability to develop further into a red blood cell, platelet, or certain kinds of white cells. A cell destined to become red blood cells transforms into a megakaryocyte-erythroid progenitor cell, proerythroblast, erythroblast, normoblast, reticulocyte, and finally an erythrocyte or mature red blood cell. Each of these intermediates has its own characteristics. For example, at the normoblast stage, the cell has a fully condensed nucleus. The most dramatic changes involve the loss of the nucleus and other cell organelles. Reticulocytes have lost the nucleus and most other cell organelles, but have not yet fully assumed a biconcave shape. Throughout the maturation process, the cell continues to produce hemoglobin. Finally, a mature red blood cell begins to transport oxygen.

"White blood cells" is a catchall term that includes many different types of cells of the immune system. The main groups of white cells are granulocytes, lymphocytes, and monocytes, but each of these can be further divided into multiple subgroups. Granulocytes include neutrophils, eosinophils, and basophils. Lymphocytes include T cells and B cells. They make antibodies and attack pathogens and other invaders. Neutrophils kill fungi and bacteria. Eosinophils attack parasites and infection. Eosinophils, mast cells, and basophils are also involved in asthma and allergies.

Blood Types

In the early 1900s, Karl Landsteiner discovered the ABO blood types. Until that discovery, blood transfusions were risky business. When mixed, some blood combinations coagulated. Landsteiner determined that blood cells carry genetically determined antigen markers. The markers were labeled A and B, and the groups carried either (groups A and B), both (group AB), or neither (group O) (Dean, 2005a; Ewald and Sumner, 2016). The immune system of individuals who lacked the particular marker would see the other blood types as foreign and attack them. For example, group O has neither marker and can be given to anyone. For this discovery, Landsteiner received the Nobel Prize.

The other important blood marker is called the Rh factor (Dean, 2005b). If you have the factor, you are Rh positive, and if you don't, you are Rh negative. Thus, individuals who have none of the ABO or Rh factors are termed "O negative." Blood from these "universal donors" can be given to anyone. Many other blood markers are known. For tissue typing for organ transplants, the HLA antigens in the blood are typed (Deaglio et al., 2020).

Clotting

Blood clotting is something that most people take for granted. It seems simple. Blood usually clots within a few minutes. However, no matter how simple it seems, blood clotting involves a complicated cascade of molecules (e.g., clotting factors) and cells (e.g., platelets) and results in the transformation of blood from a liquid to a solid (Smith et al., 2015). It also ensures that the system can staunch the loss of blood and begin to repair itself.

Here is a short version of that complex process. It begins when the single layer of endothelial cells that line blood vessels is damaged. Platelets then interact with the exposed collagen in the damaged epithelium through their glycoprotein Ia/IIa receptors. Von Willebrand factor adds to that interaction. Once the collagen, platelets, and von Willebrand factor bind, platelet glyco-protein Ib/IX/V and the A1 domain interact. Then glycoprotein VI joins to start another cascade that activates platelet integrins. The activated platelets release ADP, serotonin, platelet-activating factor, von Willebrand factor, platelet factor 4, and thromboxane A_2. This actions activate a G_g-linked protein receptor cascade involving protein kinase C and phospholipase

A_2, which causes glycoprotein IIb/IIIa to bind to fibrinogen. The platelets change shape and become even more linked by glycoprotein IIb/IIIa.

After this series, two cascades are initiated that result in the transformation of fibrinogen to fibrin: the contact activation (or intrinsic) and the tissue factor (or extrinsic) pathway. The extrinsic pathway is the main path, but both involve serine proteases and a long list of cofactors that eventually yield fibrin. Regardless of the pathway, the final common path involves the transformation of prothrombin to thrombin. Factor VIII also crosslinks the fibrin into a network.

Numerous other molecules are involved, including calcium, phospholipids, vitamin K, protein C, antithrombin, tissue factor pathway inhibitor, plasmin, prostacyclin, and more. This extraordinarily complex system is needed to regulate a critical function. Blood must clot under specific circumstances, but clotting can and sometimes is fatal if it occurs at an inappropriate place or at the wrong time.

Clotting involves the transformation of blood from a liquid to a solid, and like any biological process, anything that can go wrong in this complex and delicately balanced process will at some point. Disorders are common (Ong et al., 2018), and they involve too little clotting and too much clotting.

Too little clotting characterizes a number of blood disorders. Hemophilia is one. In most cases, the cause is inadequate levels of factor VIII or factor IX. The incidence of hemophilia A is about one in 5,000 male live births. Because it is an X-chromosome-linked disorder, it occurs more often in males. In the past, a diagnosis of hemophilia meant a short life. However, advances in medicine have enabled hemophiliacs to live an almost normal lifestyle. The identification of coagulation factor concentrates and then, in recent decades, virus-inactivated and recombinant versions has dramatically improved the lives of hemophiliacs (Franchini and Mannucci, 2012). These prevent the bleeding and joint damage that characterize this disorder. The remaining challenges involve antibodies to the coagulation factors and the short half-life of those factors.

A more common problem is a low level of vitamin K in newborns (Araki and Shirahata, 2020). Because of this, they can suffer severe bleeding and possible brain damage. The bleeding can occur anywhere in the body and so can be difficult to diagnose. The problem might result from one of several causes, including poor transfer of vitamin K through the placenta, a low level of the vitamin in breast milk, or an immature infant gut flora that does not facilitate intestinal absorption. Newborns are routinely given an injection of vitamin K shortly after birth.

Too much clotting or clotting at the wrong time and place can also be highly detrimental. Clots can form inside blood vessels. They are extremely dangerous. A clot that blocks an artery in the heart can cause an infarction or heart attack. The area of the heart that depends on the blood supply from that artery might die from lack of oxygen. Similarly, a clot that blocks an artery in the brain may result in a stroke and the death of brain tissue. Heart attacks and strokes are major causes of death and disability.

Blood clotting is another complicated function that has evolved over time. The two pathways evolved separately in vertebrates. The tissue factor or extrinsic pathway (i.e., factors VII, IX, X, prothrombin, tissue factor, and factors V and VII) evolved about 430 million years ago. The contact or intrinsic pathway (factor XI, XII, prekallikrein) appeared about 380 million years ago. Both seem to have developed by gene duplication events. Mariz and Nery (2020) carefully analyzed the evolution of clotting and found no pattern of convergence. The processes are complex so that cetaceans and fishes are quite different even though they share the same environment.

Artificial Blood

Loss of blood is a serious medical issue, and replacing the lost blood requires blood type matching in most cases. Blood also has challenges with availability, portability, risk of contamination and shelf-life. Thus, an artificial replacement for blood has been a long-standing goal. Even the famous English architect Sir Christopher Wren, who rebuilt London after the Great Fire in 1666, also worked on synthetic blood replacement of ale, wine, opium and other fluids (Winslow, 2007). Those replacements have evolved beyond those based on red blood cells to ones based on free hemoglobin (Sen Gupta, 2019). In addition, other oxygen carriers (e.g., perfluorocarbons and iron-containing porphyrin systems) are under evaluation. To date, none has been able to replace the original.

Evolution of Blood

Blood cells have also evolved. Phagocytes are among the most primitive cells. They are found even among the sponges. For that reason, phagocytes in human blood (i.e., macrophages and neutrophils) are assumed to be the oldest cells in blood and the other cell types evolved from those (Nagahata et al., 2022).

One aspect of red blood cells might offer a hint to their evolution. They differ greatly in size by species. Jawless fish have larger red blood cells than mammals, and one might assume that large cells are a primitive feature. However, those from some amphibians are even larger. Also oxygen transport capacity and hemoglobin concentrations do not correlate with the size of the red blood cells. Snyder and Sheafor (1999) pose another hypothesis: the size of red blood cells is related to the diameter of the capillaries. Cold-blooded animals require less oxygen than warm-blooded animals, and larger capillaries are adequate for that purpose. For warm-blooded animals, the capillaries are smaller so that blood can penetrate deeper into tissues to facilitate the higher energy levels to maintain a body temperature greater than ambient.

The evolution of blood can also be traced through the hemoglobins. The most familiar globin molecules is the hemoglobin that is found in vertebrate red blood cells. Four globin proteins are combined with four hemes and one iron. Hemoglobin binds oxygen and is found in nearly all vertebrates. Each molecule of hemoglobin contains four molecules of the protein globin. The description of hemoglobin is a bit technical. The key components are an iron ion that is held in place by four nitrogens. This assembly reversibly binds two atoms of oxygen for transport. Globin itself is a globular protein of about 65,000 daltons. It is water soluble and unglycosylated (it lacks the carbohydrate moieties that many proteins contain).

We are accustomed to seeing blood as red. Our blood is red because hemoglobin contains iron and is red in color. When we look at our veins through our skin, they look blue. but that isn't the case in all animals. Mollusks (e.g., cephalopods, gastropods) and most arthropods have blue blood (Burmester, 2001). Their blood is blue because they use a copper-containing molecule called hemocyanin to transport oxygen rather than the iron-containing hemoglobin used by mammals. Deoxygenated hemocyanin is colorless, but it is very blue when oxygenated. In addition, unlike hemoglobin, hemocyanin is not found inside a cell. It is in solution in the blood. The blue blood of the horseshoe crab (order Xiphosura) is quite valuable. Their blood contains amebocytes, which are similar to white blood cells in humans. A lysate of those cells, Limulus amebocyte lysate, is used as a component of a laboratory test for endotoxins. Annelid worms and marine polychaetes use the protein chlorocruorin to transport oxygen in their greenish blood. Other marine invertebrates (e.g., sipunculids, priapulids, brachiopods) use hemerythrin, which is violet-pink when oxygenated. Hemovanadin is particularly unusual. It is green, and the blood of sea squirts is light yellow. The metal in hemovanadin is vanadium, but it does not carry oxygen.

Methemoglobin is another form of hemoglobin. The iron ion in normal hemoglobin is in the Fe^{+2} state. In methemoglobin, it is Fe^{+3}, and in that state, it cannot bind oxygen. It is blue-brown in color. Enzymes can cause methemoglobin to switch back to hemoglobin. Normally, 1%–2% of human hemoglobin is methemoglobin and results in no problems. Various environmental or genetic conditions result in higher levels that cause disorders.

Many vertebrate and invertebrate animals use hemoglobins to transport oxygen, and the studies of the genes involved have yielded insights into the evolution of those proteins and species (Hardison, 2012). The genetics involve translocations and duplications. However, there are other globins. Myoglobin is has one globin, one heme, and one iron molecule. It is found mostly in muscle tissues and moves oxygen to the mitochondria. It evolved to supply oxygen to meet the environmental demands (Rummer et al., 2013). Cytoglobin is found in multiple tissues, and neuroglobin is found in brain and other tissues. It is similar to neuroglobins found in invertebrates. The hemoglobins are a fascinating group of proteins. Their genetics, regulation, and evolution provide insights into our own evolution and development.

The most amazing aspect of evolution of a system is just how the various parts came together to result in the function or tissue. While the assumption for some time has been that complex molecules develop slowly, that might not always be the case. Hemoglobin is a good example. It appeared just before the jawed fish separated from the rest of the vertebrates more than 400 million years ago. Pillai et al. (2020) showed that hemoglobin began as a monomer and later developed α- and β-subunits by gene duplication. Eventually, the dimers formed tetramer hemoglobin. These developments required only two mutations.

Heart

Any discussion of blood needs to also include the organ that allows it to do its main job: the heart.

The heart is the first organ to develop in the growing embryo because all of the other tissues and organs depend on the oxygen and nutrients in the blood that the heart pumps to all parts of the embryo. The human heart is a little larger than a person's fist. It beats 60–100 times a minute. It can beat a lot more, depending on our activity. The numbers are staggering. In a day, the heart beats about 100,000 times. In a year, it is 35 million times. And amazingly, in the average lifetime, it beats 2.5 billion times. The heart rests between beats.

Heart cells, and specifically cardiomyotes that contract, beat in unison. Isolated cardiomyocytes in a culture dish begin contracting or beating on their own. However, if they are in contact with other cardiomyocytes, they will all begin beating in synchrony (Hayashi et al., 2017; Sakamoto et al., 2021). It is quite a sight, but it underscores the importance of the flow of blood. The heart is the first organ to develop in early embryos. The other organs and tissues need a steady supply of blood with its oxygen and nutrients so that they can survive and grow.

In 1628, English physician William Harvey published his study called *De Motu Cordis*. It was the first accurate description of the circulatory system. Our hearts have four chambers: two atria and two ventricles. The right atrium receives deoxygenated blood from the body and passes it to the right ventricle, which pumps it to the lungs. The left atrium receives freshly oxygenated blood from the lungs and passes it on to the left ventricle, which pumps the blood on to the rest of the body. Arteries move the blood to the distant parts of our body. They get smaller and smaller until the blood passes into the capillaries and then onto the veins that return the blood to the heart. The capillaries are tiny and penetrate deep into tissues so that oxygen can then diffuse to those tissues.

Obtaining oxygen for internal tissues is one of the major evolutionary problems, and different animals have solved that problem with different solutions (Monahan-Earley et al., 2013). Some animals with only two tissue layers (e.g., sponges) have a central cavity. Water is pulled through pores in the sides of the animal, and through the cavity. The cells are close enough to the pores and water flow so that oxygen can diffuse to them. Some animals with three tissue layers (e.g., flatworms) have no internal system for transporting oxygen. However, they have low rates of metabolism and small, flat bodies, and simple diffusion is adequate to provide sufficient oxygen to make energy. They are referred to as *acoelomates*. Larger and more complex animals need greater amounts of oxygen than can be supplied by diffusion, and they developed systems to transport oxygen internally. Among those animals are (1) those with a coelom, (2) a vascular system, or (3) both.

A coelom is a body cavity between the outer body wall and the digestive tube. The coelom is lined with a type of epithelial cells called mesothelium and filled with fluid. No pumping system is used. Cilia on the mesothelial cells and body wall contractions mix the fluid. In contrast, vascular systems contain blood in vessels. Invertebrate vessels are lined with only a matrix. Vertebrate vessels are lined with endothelium. The blood is circulated by contractions of contractile cells in the vessel lining or by a muscular pump, the heart. More sophisticated circulatory systems also have many additional functions, including clotting, immune functions, filtration systems, and more. The vascular systems can be open or closed. Invertebrates, other than those with a coelom or relying on diffusion, have an open system. The open system refers to fact that the blood flows into a hemocoel that is in contact with the organs. Organs farther away from the heart get less oxygen. A vertebrate closed system contains the blood inside vessels that is moved by contractions of the heart.

Information on our physiological systems can come from unusual places. Some organisms, separated evolutionarily by hundreds of millions of years, sometimes have homologous genes for similar functions. These highly conserved systems underscore the importance of those systems. For example, Rosental et al. (2018) showed that our blood and immune systems share similar genes with those of the colonial ascidian tunicate, stat tunicate (*Botryllus schlosseri*). These fascinating animals begin as individuals that look something like tiny tadpoles. Eventually, they clump together to form a structure that looks like a bunch of flowers, but each petal is a single animal. However, although they are individuals, they share blood and immune cells freely. They also have a blood stem cell niche that functions similarly to the bone marrow in mammals. Finally, they have a gene that encodes the protein BHF, which is used to determine whether an individual is compatible with the other animals in a group. If so, they can then allow their blood systems to be joined. This system is similar to the determination of self and nonself in mammalian immune systems.

Fossils of soft tissues are notoriously hard to find, but researchers have had some great successes in recent years. As one example, Ma et al. (2014) found what is, so far, the oldest example of a heart and circulatory system in an arthropod fossil that is 520 million years old. It shows a well-developed system of vessels that supply the head and abdomen with a tubular heart in the back of the animal. That occurred about 600 million years ago. Endothelium improves the flow of blood and provides a good barrier between the moving blood and the tissues. It appeared about 510–540 million years ago.

Birds and mammals are warm-blooded. Maintaining a constant body temperature that is often higher than the ambient temperatures requires energy. The trade-off is that warm-blooded animals are able to move faster to obtain sufficient resources to maintain a higher body temperature. The animals achieve this by having two separate circulatory systems. A low-pressure system circulates blood to and from the lungs, and a high-pressure system pumps oxygenated blood to the rest of the body.

Cold-blooded amphibians have a three-chambered heart. In the three-chambered heart of a cold-blooded frog, the right side of the heart receives deoxygenated blood from the body, and the left side get oxygenated blood from the lungs. The single ventricle receives blood from both and pumps out a mixture of the two to the rest of the body.

Reptiles are somewhere in between, and the evolutionary change from a three- to a four-chambered heart was a significant question in biology. For example, turtles feature a single ventricle, but a septum is beginning to form to separate the ventricle into two chambers. The result is blood that is better oxygenated than that in amphibians and that will support greater levels of activity (Figure 10.1). A better understanding of the development of the septum was obtained by the laboratory of Benoit Bruneau (Bruneau, 2009). They study a protein called Tbx5. Tbx5 is a transcription factor, which means that it regulates the expression of multiple genes. It turns some on and others off at specific developmental times. They found that Tbx5 was turned on throughout the hearts of amphibians (Koshiba-Takeuchi et al., 2009), but in warm-blooded animals, its expression was limited to the left side of the heart so that it allowed the development of a septum that separates the ventricle into two chambers. In turtles, they found an intermediate situation with expression on the left side that diminished into the right side. This gradient promoted the development of a partial septum.

Lungs

We human adults breathe 8–16 times per minute, and each breath is about 500 mL. Infants breathe a lot faster: up to 44 times a minute. What happens to that air? Inhaled air passes through our nose and/or mouth and down our trachea and into our two lungs. The air passes through the primary, secondary, and tertiary bronchi and into the bronchioles and finally to small sacs called alveoli (Travaglini et al., 2020). There are about 240 million of

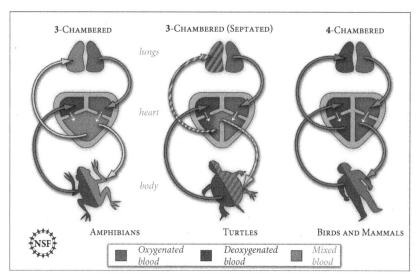

Figure 10.1. Evolution of reptile hearts. Frogs have three-chambered hearts. Oxygenated and deoxygenated blood are mixed in the ventricle, and so, frogs do well with partially oxygenated blood. Turtles have blood with more oxygen due to a partial septum that partly separates the ventricles. Birds and mammals have two separate circulatory systems. Blood is pumped at low pressure to the lungs and at high pressure to the rest of the body. The blood is fully oxygenated. Zina Deretsky, U.S. National Science Foundation.

them in each lung (Ochs et al., 2004). In the alveoli, a very thin membrane separates the gaseous phase from the blood in the pulmonary veins. The right lung has three lobes, and the left lobe has two.

We do not have to think about breathing. Signals from the respiratory center in the brainstem enervate the diaphragm. The base of the lung sits on the diaphragm, a muscle in the chest. As the diaphragm contracts, it creates a negative pressure in the lungs that draws in air. When it relaxes, it returns to its original position, and we exhale. Other muscles assist in these actions, especially when we are engaged in active exertions.

Breathing is also controlled by the autonomic nervous system. The vagus nerve transmits signals from the parasympathetic nervous system to cause the smooth muscle in the bronchus to contract and increase the production of secretions in the lungs. Other signals cause bronchodilation.

Oxygen is essential to most life forms on Earth today, but it is also a toxic substance that must be carefully regulated (Hsia et al., 2013). Multiple systems have evolved to take advantage of oxygen and minimize its damage and eliminate carbon dioxide. These systems began as simple gas diffusion surfaces, but then evolved to more sophisticated organs.

The very first organisms on Earth were anaerobes. Oxygen was a problem for them, and they must have developed antioxidant defenses against oxygen and reactive oxygen species. Metalloproteins serve these functions well. Metals are commonly used in enzymes. Today hemoglobin and chlorophyll are examples of such molecules that use iron and magnesium atoms, respectively. Metalloproteins have been around since the last universal common ancestor that connected bacteria, archaea, and eukaryotes.

The atmosphere changed as cyanobacteria developed photosynthesis that released oxygen to the environment. By about 500 million years ago, the amount of oxygen reached something like modern-day levels. Those levels facilitated the transition of aquatic organism to move onto land. Also, water contains only 3% of the oxygen of an equal volume of air. So it is much more energy efficient to obtain oxygen from the air.

While the oxygen concentrations were high, land and flying arthropods developed mechanisms to obtain oxygen by diffusion across their exoskeletons. About 200 to 240 million years ago, the oxygen concentrations dropped for unknown reasons, and many insects and other land animals died. Those that survived developed new systems to obtain oxygen. Those systems included endothermy and separation of oxygenated and deoxygenated blood. The oxygen levels slowly recovered, and that allowed the evolution of vertebrates, including reptiles, mammals, and birds.

Invertebrates have many organs for handling gas exchange, and those organs are quite different from species to species, according to their particular needs. Those organs also resemble the ones found in the fossil record. The few systems covered here in broad terms are meant to provide examples of early gas exchange mechanisms. Snails are thought to have developed the first gill-like organs. Some have a lung siphon that allows them to draw in air and to remain underwater for up to an hour. Others combine diffusion through their body with a primitive lung. Some use hemocyanin to transport oxygen, and others use hemoglobin.

Crustaceans use a variety of gas exchange strategies to oxygenate hemocyanin. Some use gills to breathe air. Woodlice (*Hemilepistus* sp.) use gills and lungs and tracheae. Still others use just lungs or tracheae. Yet others use lungs on the pleopods (i.e., the small swimmerets behind the main legs). Decapoda use multiple systems. Aquatic or semiaquatic crabs use their gills to breathe air. Land crabs have lungs and gills and can switch between the two as needed. Some species supplement their gas exchange with diffusion through thinned areas of the cuticle on their legs and other body areas. Many aquatic and land crabs use a pumping organ called a scaphognatite for active ventilation. Arthropods, such as spiders, have book lungs. These structures have stacks that alternate between air pockets and tissue with hemolymph. They look like a book, hence the name.

Many other arthropods use a combination of spiracles (i.e., openings in the exoskeleton), trachea, and diffusion of gases along with active or inactive ventilation. For example, arachnids use only diffusion. Scorpions were probably the first arachnids to evolve, and they used simple book lungs that were probably also the first air-breathing organs. Some early arachnids used only tracheae and spiracles. Most spiders have two pairs of book lungs just behind the thorax and tracheae. Interestingly, these four-lunged spiders only breathe at rest. During periods of activity, they are anaerobic. Spiders have an open circulatory system, and the four-lunged spiders have separate anterior and posterior circulations. Insects also have an unusual breathing pattern. Slow-moving insects, such as velvet worms, breathe continuously. But active insects, such as flying insects, breathe discontinuously and can lower their metabolic rate to almost zero. As noted above, insects have spiracles with valves that lead into tracheae. The tracheae penetrate deep into the tissues so that the diffusion is minimal.

Interfilament water channels were a major development in the gills of fish. They allowed water to be sucked into adjacent lamellae by water exiting gill orifices where the water had already been depleted of oxygen. This offers advantages. Branchial constrictor muscles are another advance that was further improved in fish that evolved later. The bony fishes (Osteichthyes) have even more sophisticated gill systems. Many fish have also evolved structures that allow them to breathe air. In some of these, swim bladders are used, and branched organs extend deep into the other structures of the fish.

The swim bladder might have been the predecessor of vertebrate lungs, but this assertion is controversial. Did lungs become the swim bladder or did the swim bladder come from lungs or was it a completely different mechanism? These questions are not yet answered. The one thing that is clear is that lungs appeared with the amphibians and they have been further refined since then.

Evolution moves in strange ways. Structure begin with a function, but that function can change over time, depending on the needs of the organism.

For example, insect wings likely began as gills in aquatic arthropods. For example, Damen et al. (2002) showed that two genes (i.e., *pdm/nubbin* and *apterous*) associated with the development of insect wings are also involved in crustacean gills. In spiders, the same genes help with the development of book lungs, tracheae, and spinnerts that are used in breathing. These changes were initiated by the movement of organisms from an aquatic to a terrestrial existence. Daniels et al. (2004) suggest that lungs could not have succeeded with the evolution of surfactant. Surfactant allows lungs to work effectively by decreasing the surface tension between the gas and liquid interface in the lungs.

Kidneys

Kidneys are also remarkable organs. They are shaped like beans, are reddish-brown, and located on the right and left sides about halfway down the abdomen and toward the rear. Each is about 12 cm long. The kidneys for most vertebrates are similar, and so, the beef or lamb kidneys in the meat counter at the supermarket will provide a good idea of what human kidneys look like.

The kidneys cleanse the blood of waste products, which are routed into the ureter with the urine and carried to the bladder and then out through the urethra. The functional unit is the nephron, and human kidneys have about 1 million of them. The kidneys control the level of hydration. Excess water is eliminated in the urine, and urination can be limited when water is short. Kidneys reabsorb water, sodium, bicarbonate, glucose, and amino acids. They eliminate ammonium, potassium, uric acid, and urea.

Urine also contains hormones, proteins, and various metabolites. The normal color of urine is light yellow, due to urobilin, the final breakdown product of heme protein from worn-out red blood cells. Color changes are normal, but excessive changes to color or composition can indicate disease states.

One example is the smell in urine after a person eats asparagus (*Asparagus officinalis*). The smell comes from various metabolites of the sulfur-containing molecule asparagusic acid. It is still unclear whether everyone makes the molecule or whether some people simply cannot smell it. However, Eriksson et al. (2010) reported a single-nucleotide polymorphism in a region of chromosome 1 that is associated with olfactory receptors. A few crowd-sourced studies of people recruited to eat asparagus and smell urine have been published, but have not solved the problem yet. The debate continues.

Kidney disease is a serious and growing problem (El Nahas and Bello, 2005). As our population ages overall, the incidence of kidney disease also increases. The only treatments for end-stage renal disease are kidney transplantation and dialysis. Both are expensive. Urinary tract infections are common ailments, and kidney stones are also not uncommon.

The evolution of the kidneys is quite fascinating. Life first appeared in the sea. In fact, life spent the first billion years or more in the sea before venturing onto land. Even then, the first organisms on land were likely single cells. Multicellular organisms did not appear until about 600 million years ago. So living organisms spent a very long time in the sea, and they adapted well to that environment. Those living creatures first had to stabilize their internal environment. The extracellular fluid of all living things is relatively similar in terms of ions (e.g., sodium, potassium, chloride, and calcium). The early fish in those primordial seas had the challenge of maintaining that osmolarity in an environment (i.e., sea water) that had a much higher concentration of salts. Simple diffusion causes water to move to areas of higher ionic concentration, and so, the fish had to deal with the constant loss of water. To accomplish this, they drank a lot of sea water to replace the water lost. That brought on a second problem of what to do with the excess salts that came with the sea water. They solved that problem with something like a primitive kidney: an energy-dependent ion pump in the gill epithelia. The kidney appeared later.

So that solved the problem for saltwater fish. However, fish then moved into freshwater rivers and faced the opposite problem. Freshwater ion concentrations are much lower than those in the organism. Those fish refined the kidney to enhance the glomerular filtration rate so that they excrete urine with less salt than the intracellular concentrations.

The evolution of the kidney includes three organs. The pronephros is the first and the one still used by some amphibians and some lower fish. It has a single nephron that is attached to a nephric duct and then to the cloaca. The more sophisticated mesonephros is used by most fish and amphibians. These first two organs appear briefly in the development of human embryos, but are essentially discarded early on. Finally, the metanephros is the functional kidney of land vertebrates.

Invertebrates evolved excretory systems too (Mahasen, 2016), and they generally feature one of three organs. Protozoa have a contractile vacuole that is involved in managing the water and salt balance. They deal with nitrogenous wastes by diffusion. Water is added to the vacuole until it reaches a certain size. Then the vacuole moves to the plasma membrane and the excess water is drained through a small pore. Most invertebrates have one or the other of two types of nephridia. It's also called a flame cell system. Some, such as flatworms, have a protonephridium. The nephridium is a tubule with a ciliated flame cell at the end. Earthworms and other invertebrates have a metanephridium that features an open tubule system, a vascular network, and a coelom at one end and opens into an excretory pore at the other. Arthropods have Malpighian tubules and rectal glands. There is no interface with the blood system. Body fluids move into the tubules by osmosis and eventually are reabsorbed by the rectal glands.

Internal Sea

Life on Earth began in the sea, and many of our systems are attempts to maintain that water-based environment, even on dry land. Single-celled organisms depend on simple diffusion or other strategies to import nutrients and export waste products. Multicellular organisms grew larger and needed systems to transport nutrients, facilitate gas exchange, and eliminate wastes. Those systems usually serve us well. However, nothing in biology is perfect, and most of us have experienced minor problems. A small cut or abrasion clots and heals in a short time. We cough to clear our lungs during common infections, such as influenza or colds. The heart, lungs, and kidneys are intimate components of the circulatory system and are among those most involved with maintaining that aqueous environment.

11

Frogs and Fingers

The Very Hungry Caterpillar (Eric Carle, World of Eric Carle, 1994) tells a story that is well-known to most school children. An egg on a leaf hatches to release a caterpillar. The caterpillar eats and grows much bigger. After a time, it attaches itself to the underside of a leaf and surrounds itself with a cocoon. The pupa remains within the cocoon for a couple of weeks and then emerges as an adult butterfly. This is a simple children's book, but it provides a surprisingly good explanation of the four stages of complete insect metamorphosis: egg, larva, pupa, and adult.

Metamorphosis of butterflies is an amazing event. The great magician Houdini copied it. He named one of his most famous tricks Metamorphosis. The magician's assistant is locked in a box. The magician climbs onto the box and pulls a curtain up around the box and himself. On a count of three the magician pulls the curtain up to hide himself, and immediately, the curtain is dropped to reveal the magician's assistant. In nature, metamorphosis is more complicated, but in the end, a caterpillar is transformed into a butterfly. To make this dramatic change work, whole tissues are rearranged. Many cells die, and new cells are produced.

The real magic of metamorphosis is programmed cell death. As far back as the 19th century, biologists noticed that lots of cells were lost during the development of some organisms. In the mid-1960s, Richard Lockshin and Carroll Williams at Harvard first used the term "programmed cell death" to describe development in the American silkmoth (Lockshin, 2016). They noted that the moth's abdominal muscle cells died during development. The larvae use those muscles to move, but they are not needed in the adult. Their last function is to push the moth's blood (called hemolymph) into the wing veins to make the wings expand to their full size. After that, the muscle cells die.

While the name is disappointingly mundane, programmed cell death is one of the most extraordinary developments in evolution (Ameisen, 2002). As the term implies, cells are programmed to die. This means that there are programs of genes that, once activated, result in the death of the cell. The cell

The Biology of Us. Gary C. Howard, Oxford University Press. © Oxford University Press 2024.
DOI: 10.1093/oso/9780197664797.003.0011

commits suicide. These molecular processes cause the colors of the leaves to change and much more. Fruit ripens and falls to the ground. Tadpoles lose their tails, grow legs, and change into frogs. And caterpillars become butterflies. Programmed cell death is critical to human development and health. While that thought might be counterintuitive or even disconcerting, it is the real magic. Living cells have grabbed onto death and now use it as a tool in development and to defend against disease. In fact, without it, we humans could not survive. It gives us fingers and a working immune system and much more.

In living organisms, anything that can go wrong eventually does. Thus, organisms have evolved mechanisms to handle those emergencies. Damaged or worn-out proteins and organelles are dismantled and recycled. Enzymes are available to repair damage to DNA. Even whole cells can go bad. They can become infected by bacteria or viruses that can spread to other cells. Cells can lose their ability to control their own growth. That might result in them overgrowing the limits of a particular tissue or organ. It might also indicate that they have become malignant. In either case, they pose a threat to the rest of the organism. Finally, the cell might be just fine, but has become unnecessary or even in the way as the organism enters a new phase of development. Fortunately, cells have also evolved programmed cell death to handle these situations.

Programmed cell death (Ameisen, 2002) involves multiple mechanisms that will be described in detail below. They include apoptosis, autophagy, necroptosis, ferroptosis, and several more. They also involve the enzymes that accomplish the specific chain of events that define each type of cell death. The molecules that initiate programmed cell death are held in check by other genes and proteins in a delicate balance. Finally, these genes are found in every organism studied so far and across many phyla and divisions. The death genes and pathways are conserved from worms to humans (Baehrecke, 2003). They have been found in essentially every cell that has been examined, including all human cells. In addition, the genes are very similar in a wide variety of organisms, and that suggests that those genes evolved long ago and are very important to the organisms.

The benefits of programmed cell death are not hard to envision. Evolution operates by rewarding changes or mutations that give one organism an advantage over others. Those organisms reproduce more effectively than those without the mutation, and thus, they outgrow the others. Multicellular organisms are enormous collections of cells that must be carefully coordinated in their functions. This requires considerable organization to ensure that the cells, tissues, and organs are correctly positioned and functioning. If some cells go bad, the organism needs to eliminate them and initiate appropriate repairs. Mutations that sacrifice some cells so that the whole organism can survive would be an evolutionary advantage. Programmed cell death is particularly valuable to multicellular organisms. It allows them to deal with infections and cells that have gone rogue. It also provides a mechanism to eliminate cells that are no longer needed at a specific developmental stage. Thus, programmed cell death seems to have conferred an advantage on organisms, and that strategy was taken up by many (maybe all) organisms long ago.

We have all seen examples of programmed cell death, but we might not have realized the magic. Yet, they are even more magical when you know a bit about the mechanisms that make them possible. In this chapter, we will look at how that happens. We begin by describing programmed cell death and its several mechanisms. We will also look at its important workings in different organisms across the spectrum of living organisms, including humans.

Programmed Cell Death

Many cells have a normal "life span" after which they simply wear out, die, and are discarded. Some cells, especially those that are exposed to the external world (e.g., skin cells), wear out with that exposure and must be replaced. Most organisms, including humans, have analogous sophisticated systems to remove dead cells. Those systems are very important for maintaining health, and they are well used. Hundreds of billions of cells die every day in a human body. These cells are damaged by normal physiological processes or interactions with pathogens. There are multiple types of programmed cell death, and some feature characteristics that overlap with other forms (Ketelut-Carneiro and Fitzgerald, 2022).

Apoptosis

Apoptosis is one of the most important forms of programmed cell death (Kerr et al., 1972). It is defined by a set of specific characteristics. The cell shrinks in size. The chromatin, which contains the DNA and genes, condenses and fragments. "Blebs" or bubble-like protrusions form on the cell membrane. Finally, the cell breaks apart into small apoptotic bodies, which contain cellular material and are bounded by fragments of the cell membrane. While apoptosis is primarily used to eliminate diseased, damaged, infected, or unneeded cells, it can also be used to maintain cells. For example, caspase-8 is important for the well-being of adult T cells, and caspase-3 is used in tissue differentiation, regeneration, and neural development (Shalini et al., 2015).

Autophagy

Autophagy is a key element of many processes and disorders (Khandia et al., 2019). The term "autophagy" is a combination of "auto," which means self, and "phagy," which refers to eating. Three types of autophagy are known (Tsujimoto and Shimizu, 2005; Shibutani et al., 2015). Chaperone-mediated autophagy uses specific receptors to transport individual proteins into the lysosome, where they are degraded by the lysosomal enzymes. Chaperones are proteins that bind other specific proteins and escort them to the receptor on the lysosome membrane. Microautophagy captures cellular components by invaginating the lysosomal membrane. Once the membrane surrounds the material, it pinches off to form a vesicle inside the lysosome. Lysosomal enzymes then quickly degrade the vesicle and its contents. Macroautophagy (hereafter referred to as autophagy) defends the cell against pathogens and regulates immune responses.

Autophagy is highly regulated at multiple levels, including at the gene, RNA, and protein levels. In yeast, 38 autophagy genes are known, and 15 of those are also found in humans. Scientists say that those 15 genes are conserved from yeast to humans. Autophagy becomes less effective with age, and expression of the autophagy genes decreases with aging. In turn, damaged cellular components accumulate. Multiple factors regulate autophagy, but the main one is called mTOR (mammalian target of rapamycin). mTOR is a major regulator in cells and controls many processes involving metabolism. When conditions are favorable for growth, it encourages growth and inhibits autophagy. When nutrients are lacking, autophagy is used to degrade and recycle the cell's own components. Autophagy is initiated when a membrane-bound phagophore forms from the endoplasmic reticulum. It senses targets for degradation and moves to capture them. The membrane closes around the target to form a double-membrane-bound autophagosome. The autophagosome is directed to the lysosome by microtubules. Its outer membrane fuses with the membrane of a lysosome and releases its contents into the lysosome. The lysosome environment is acidic and contains degradative enzymes that digest the cargo. Once the cargo is degraded, the resulting raw materials are recycled by the cell.

Necrosis

Necrosis is not a form of programmed cell death because it is less "programmed" than apoptosis (Kroemer et al., 2016). Necrosis is seen in trauma, infections, cancer, and infarctions. The cause might include a loss of blood supply (e.g., infarction, ischemia), frostbite, or spider and snake venoms that contain enzymes that kill cells. Necrosis is not completely unregulated, and it may occasionally share some features with apoptosis. Yet, they are distinct processes. In necrosis, the plasma membrane is damaged, and water enters the cell. The cell swells, and its contents leak out. Degradative enzymes in lysosomes that are integral to autophagy escape and add to the cellular damage. Also, signaling molecules set off an immune reaction. Phagocytes are attracted to clean up the debris. However, they also release molecules that damage surrounding cells and complicate the healing process by adding to the number of dying cells. Gangrene may result.

Necroptosis

Necroptosis seems to be in the middle of a spectrum between apoptosis and necrosis (Berghe et al., 2010). It resembles necrosis but is more organized. Like necrosis, necroptosis is independent of caspases. However, the breakdown of the cell membrane is controlled. Several factors combine to form a complex called the necrosome. The membrane begins to fail, and the leakage attracts immune cells to clean up the debris. Sometimes for unknown reasons, apoptosis fails, and that might be a signal for necroptosis to begin. Like necrosis, necroptosis is associated with tissue damage, such as myocardial infarction (heart attack) and stroke, and also other diseases, such as atherosclerosis, inflammatory bowel disease, neurodegeneration, and cancer.

Other Forms of Programmed Cell Death

Several other forms are often associated with specific tissues or conditions. *Pyroptosis* is involved in responses to pathogens, heart attacks, and HIV infection (Doitsh et al., 2014). The cell swells until it breaks open to release the cell contents, which induce an immune response and inflammation. Caspase-1, a protease mostly known as interleukin-1β-converting enzyme, is a key feature of pyroptosis. *Cornification* is essentially the final step in the differentiation of the outer protective skin layer. Proteins are cross-linked by transglutaminases, and lipids are released into the extracellular space to insulate the body from the environment. *Anoikis* helps to maintain the appropriate size and shape of organs and tissues (Gilmore, 2005). It removes excess or mislocated cells. The integrin family is intimately involved in anoikis. *Excitotoxicity* occurs when neurons are exposed to excitatory amino acids (e.g., glutamate) and cytosolic Ca^{2+} goes out of control. It may be important in Alzheimer's disease.

Butterflies and Frogs

Metamorphosis is the transformation of an organism from a juvenile to an adult form. It occurs in insects, frogs and toads, fish, newts and salamanders, and other organisms (Holstein and Laudet, 2014). During the transformation, cells are lost, and apoptosis is a key mechanism in the process (Wittig et al., 2011), but there are others. The cell death depends on caspases.

Insects

Many insects also undergo metamorphosis (Truman, 2019; Hall and Daniel, 2019). Beetles, flies, ants, butterflies, and moths undergo what is called complete metamorphosis. Crickets, grasshoppers, and dragonflies undergo incomplete metamorphosis. In each path, the insects hatch from eggs and go through stages in which their tissues and organs change dramatically. The processes are mostly under the control of hormones. The ecdysteroids induce metamorphosis, and the juvenile hormones inhibit it. Programmed cell death is again an important part of both process. Truman (2019) asserts that the ability of insects to fly was critical in the evolution of metamorphosis.

Complete metamorphosis comprises four stages: egg, larva, pupa, and adult (Tettamanti and Casartelli, 2019). An egg on a leaf hatches to reveal a larva that we commonly call a caterpillar. Caterpillars consume several times their own body weight each day, and they grow rapidly. After a period of growth, the larva attaches itself to the underside of a leaf and forms a hard shell or cocoon around itself. The pupa or chrysalis remains within the cocoon for a couple of weeks. After a series of developmental steps, the pupa is transformed into the adult form, and a beautiful butterfly emerges from the cocoon. This event is called eclosion. The butterfly typically hangs upside down from the cocoon until its wings dry.

As noted, the term "programmed cell death" was first used to describe the loss of tissue in moths during metamorphosis (Lockshin and Williams, 1965). Cell death is deeply involved in complete metamorphosis, especially in the transformation of the larva into an adult. Nearly 80% of insects have some form of complete metamorphosis, but they vary in the extent of programmed cell death (Rolff et al., 2019). Organisms use autophagy or apoptosis to remove cells and recycle the components. Apoptosis usually occurs earlier than autophagy. In apoptosis, phagocytic cells recycle the components of the dead cell, but in autophagy, that task is accomplished by the cell itself.

Incomplete metamorphosis has only three stages: egg, nymph, and adult. The nymph is a smaller version of the adult. It lives in the same place as the adult and eats the same food. However, it lacks wings and cannot reproduce. Like all arthropods, the nymph must shed its exoskeleton to grow. Each molt is called an instar. Thus, from hatching to the first molt, the nymph is a first instar. After the first molt, it is a second instar. The number of instars vary according to the species and availability of food.

Frogs and Toads

Metamorphosis in amphibians is well known. Frogs and most toads lay their fertilized eggs in water. Unlike the eggs of reptiles or birds, amphibian eggs lack a hard shell. So the eggs must be laid in water to keep from drying out. The eggs hatch to release tadpoles. In some parts of the United States, every calm body of water seems to have tadpoles at the end of April or early May. Tadpoles are the larval stage for frogs and toads. They live in the water and breathe through gills. The tadpoles lack legs, but they swim with the aid of a long tail. For the first few days after hatching, they continue to consume the yolk from the egg. Thereafter, most are herbivores, but some are omnivores. In fact, they are even sometimes cannibalistic on other tadpoles. Their metamorphosis involves extensive changes in tissues and organs that, in the end, allow them to live on land at least part of the time (Ishizuya-Oka et al., 2010). The keys to the transformation are thyroid hormone and corticosteroids (Shibata et al., 2021; Paul et al., 2022), but thyroid hormone dominates the process. Interestingly, it induces apoptosis in the larval intestine and, at the same time, induces proliferation of the adult intestine. Another factor is clearly involved, but more research will be needed to sort this out.

Apoptosis is critical to the process, since many juvenile organs and tissues must be replaced (Ishizuya-Oka et al., 2010). Some evidence also supports a small role for autophagy, particularly in the removal of the tail muscles (Lockshin and Zakeri, 2004). The process is highly regulated, and the timing ensures that larval and adult tissues appear and disappear on schedule (Ishizuya-Oka et al., 2010). The changes begin with the appearance of the back and then the front legs. The tail disappears. Lungs begin to replace the gills. Adult frogs and toads are carnivores, and the changes reflect the new diet. The mouth grows much larger, and the intestines transform to digest a carnivorous diet. The change from newly hatched tadpole to adult frog takes about 14 weeks. When the tadpoles are about 4 weeks old, they grow teeth, and their intestines are ready to digest small insects. In week 5, back legs will start to appear. By week 12, the tadpole looks like a small frog but still has a tail. Also the lungs are functioning, and the baby frog can leave the water. At week 14, the tail is gone. Metamorphosis is complete, and the frog is a young adult.

Falling Leaves and Fruit

In the fall, the leaves on deciduous trees change colors and then fall off. Programmed cell death is behind these processes too. The scientific name for these events is abscission, and it has three steps that are carefully choreographed and timed. The first step involves the pigments (e.g., chlorophyll, xanthophyll, carotenoid, anthocyanin) that provide the colors of the leaves. Chlorophyll is valuable to the plants. It's the main pigment involved in photosynthesis, but it also contains nitrogen, which the plant uses to make amino acids, nucleotides, and other compounds. In the autumn, the amount of sunlight decreases, and the plant begins to recycle the chlorophyll and to reallocate the raw materials to other needs. The levels of chlorophyll drop faster than those of other pigments, and we see that as the colors changing from green to red, yellow, gold, and brown. The second step involves the preparation for the separation of the leaf. This occurs at the base of the petiole, which is the stalk that joins the leaf to the stem. A separation or abscission zone forms with weak cells in the top layer. Just below that, a protective layer forms. The cells release suberin and lignin to form a waterproof seal where the break will occur. In the final step, the weak cells in the abscission zone import water, swell, and burst. What initiates this step is not known, but there are several possibilities. Reactive oxygen species form naturally during metabolism, and various stresses, such as cooler temperatures, attacks by insects, or pathogens, cause their levels to increase. The plant hormones auxin and ethylene may also have a role. As levels of auxin decrease, the cells become more sensitive to ethylene, and ethylene activates the enzymes cellulase and polygalacturonase that degrade plant cell walls.

A similar series of steps results in ripe fruit detaching from the plant and falling to the ground. As the layer of cells dies, the fruit loosens and eventually falls.

Human Development and Maintenance

Humans do not undergo metamorphosis, but as we develop into adults, our tissues and organs change. In some cases (e.g., fingers and toes), extra cells are made that need to be removed. In other cases (e.g., neurons, immune), systems need to be refined. Some cells may need to be subjected to quality control (e.g., oocytes). Also the normal wear and tear on some cells (e.g., skin, blood cells, intestinal cells) means that they need to be replaced periodically. Programmed cell death is important in animal development and health (Fuchs and Steller, 2011). When it fails, the results include developmental disorders, neurodegeneration, and cancer. It uses the caspases and other enzymes to cause cell death and signals from dying cells to promote cell division, tissue regeneration, and wound healing.

Hands and Fingers

The human hand is a marvelous machine. Each hand contains 27 individual bones. Thus, our two hands account for nearly one-quarter of the total number of bones in our body. Those bones are moved by 30 muscles that are mostly in the forearm. Thin tendons connect the muscles to the fingers. Estimates of the number of nerve endings vary greatly, but one estimate is that about 15% of the 230,000 afferent fibers in the body are in the palmar surfaces of the two hands (Corniani and Saal, 2020). The bones, muscles, and tendons and their nerves and blood supply allow our hands to do an extraordinary range of things from handling heavy objects to completing delicate precision tasks.

Theories of human and ape evolution have long assumed that the last common ancestor between the two groups was a knuckle-walking creature that resembled a chimpanzee. However, more recent research has challenged that notion. The ratio of the lengths of the thumb to the other fingers is much greater in humans than in other apes. In a careful study of the hand bones of humans, apes, and fossils, Almécija et al. (2015) showed that human hands have changed little since the last common ancestor, but the hands of other apes have evolved significantly.

Our hands are simply amazing. We use them constantly, but rarely give them a second thought. They can grip with great strength, and yet, they can also perform delicate tasks. Humans have made a great investment in their hands. For example, each hand has 27 individual bones. The 54 bones of the two hands are one-quarter of the total 206 bones in the entire body. The five fingers work independently or in concert to complete various tasks. In technical terms, each finger can flex, extend, abduct, adduct, and circumduct. The fingers are exquisitely sensitive. Each fingertip has about 3,000 nerve endings, and each nerve ending senses an area of about 5 mm^2. There are two types of nerve endings. Merkel discs are found in the fingertips and the lips. They adapt slowly and respond to light touches. Meissner corpuscles are found in glabrous skin (i.e., areas that lack hair, such as the palms of the hands and feet). They adapt rapidly and detect touch, low-frequency vibrations, and flutter. The ability of our hands to sense touch may be even more amazing than originally thought. Pruszynski and Johansson (2014) reported that the tactile neurons can ascertain the orientation of edges of simple geometric features in a way that suggests a degree of information processing that would have been expected of occurring in the brain.

Our hands also may have been one of the keys to human evolution (Young, 2003). Our opposable thumb, ability with language, and bipedalism are often cited as an important characteristics that differentiated us from our ape-like ancestors. Bipedalism might have freed up the hands for additional tasks, but Young (2003) also suggested that the abilities of the human to throw objects and to hold a club might have facilitated human evolution. Pollard and colleagues found specific control sequences in the human that differ from that of chimpanzees that called these highly accelerated regions. The function of these potential regulatory regions is not clear yet, but there is some evidence that some of them control the development of the hands.

The development of our hands is a fascinating story and apoptosis is part of it (Hernandez-Martinez and Covarrubias, 2011). Our hands begin as flattened plates at the ends of the arms (Standring, 2008). The apical ectodermal ridges and mesenchyme control the development of the limbs and hands by extensive signaling with critical gene products (e.g., sonic hedgehog, Hox, bone morphogenic proteins, fibroblast growth factors). Together they induce the formation of the finger bones or phalanges. The extracellular matrix acts as a scaffold for cell aggregation, and both growth and apoptosis are involved. However, the digits are still webbed. To release the fingers, the "extra" mesenchyme between the digit rays is removed by the induction of apoptosis by retinoic acid (Suzanne and Steller, 2013). This action allows the fingers to appear as individual structures. A similar process occurs with the toes on the feet.

This process works well for humans and others, but not all animals undergo this process. Some, such as ducks and bats, lack cell death in their extremities, and the result is webbed hands, wings, or feet. In fact, the process is also not always successful in humans. Syndactyly is a condition in which apoptosis fails. As a result, the fingers remain joined by tissue, and the cause is a defect in apoptosis (Schatz et al., 2014).

Neural Development

We have a lot of cells in our brains: about 100 billion neurons and 10–50 times that number of glial cells (Herculano-Houzel, 2012). Our thoughts and memories are stored in those neurons in a manner that is not understood. What is known is that it involves connections between individual neurons that may number in the thousands. Those connections are continually being made and replaced. Although we have a lot of neurons, we have even more early on. More neurons are generated than will be needed later on, but more than half of them die during development (Barnes and Raff,

1999). The reason for the early excess and the culling is not clear. Perhaps it ensures that the correct connections are made.

Apoptosis is primarily responsible for the loss of neurons in the developing brain (Yuan and Yankner, 2000). The process is regulated by several proteins, such as apoptotic protease-activating factor 1, those of the Bcl-2 and caspase families, and neurotrophins that control protein kinase cascades. However, multiple other mechanisms are involved, including intrinsic and extrinsic apoptosis, oncosis, necroptosis, parthanatos, ferroptosis, sarmoptosis, autophagic cell death, autosis, autolysis, paraptosis, pyroptosis, phagoptosis, and mitochondrial permeability transition (Fricker et al., 2018).

Neurons are postmitotic: They no longer divide. More recently, neural stem cells have been found in the brain. So neurogenesis must continue. However, for the most part, the neurons that we are born with must last essentially for a lifetime. In fact, neurodegenerative diseases are characterized by the loss of specific types of neurons. For example, in Parkinson's disease, dopaminergic neurons die. In Alzheimer's disease, neurons in the hippocampus are lost. Also apoptosis might be the mechanism that drives the loss of neurons in those diseases.

Central Nervous System

The central nervous system (CNS) includes the brain and spinal cord. Its development begins early on in humans. A plate of ectoderm bends on two edges to form a U-shaped structure. The edges continue to expand until they meet at the top to form the neural tube. Additional mesenchymal cells migrate to position themselves underneath the neural tube. The front (anterior) part of the tube expands to form the forebrain, midbrain, and hindbrain. As this highly choreographed dance continues, more cells arrive to form other structures. Everything is carefully controlled, and apoptosis is a key element of the process. Tragically, the process does not always work correctly. The results can be as relatively simple as cleft lip or cleft palate, which can often be repaired. They can also be extremely serious. In spina bifida, the distal end of the spinal column does not develop correctly, and the fetus does not develop control of their lower extremities. In anencephaly, the skull is left open, and the brain is uncovered. This condition is not compatible with life.

Apoptosis is important in the development of the CNS. Programmed cell death and the caspases are deeply involved in neurodevelopment (e.g., axon guidance, synapse formation, axon pruning, and synaptic functions) (Nguyen et al., 2021). Interfering with apoptosis yields serious malformations (Yamaguchi and Miura, 2015). The enzymes that effectuate the cell death are caspases. They act to limit the number of cells in the various tissues and organs, such as the CNS, of the embryo.

Immune System

Our immune system is extremely sophisticated. We are surrounded by bacteria, fungi, and other microscopic invaders. In addition, our own cells can sometimes lose control of themselves and turn cancerous. The immune system evolved to deal with these problems.

The system has two main branches. The innate immune system consists of passive barriers (e.g., skin, mucous membranes, coughing, tears, stomach acid) and the innate humoral immunity that moves immediately to attack threats (e.g., complement system, interferon, interleukin-1) in a nonspecific manner. The other branch and the one that we are most concerned about here is the adaptive immunity. It comprises immune cells that directly attack the threat or that release antibodies that attack and destroy the threat. Once a threat is detected, the numbers of specific antigen-reactive T cells increase rapidly. The previously existing 50–200 T-cell precursors for a specific antigen increase 10,000- to 50,000-fold (Hedrick et al., 2010). Once the threat is neutralized, it is critical that the system go back to its "normal" state. Those extra immune cells must be eliminated (Krammer et al., 2007). Most amazingly, these cells can "remember" what the threat looks like. When the system is confronted with the same threat again, it responds more rapidly. That memory can last up to a lifetime, and it is the basis for vaccinations.

A complex and exquisitely balanced regulatory system maintains our immune system. However, it isn't perfect. It sometimes fails us by being overly aggressive and turning on our own tissues (e.g., autoimmune diseases, such as diabetes). In other cases, it is not sufficiently aggressive and allows our own cells to grow out of control (e.g., cancer). Nevertheless, it is a remarkable defensive system.

How does this extraordinary system develop? It is important to keep in mind that our immune system begins with an astoundingly immense number of receptors. Our genes encode enough information so that we can make an immune response to just about anything. How does a system with the ability to attack essentially every antigen distinguish between a legitimate threat and our own tissues? The answer is "clonal selection," and it was suggested by Frank Macfarlane Burnet and later proved experimentally by Peter Medawar. Programmed cell death is intimately involved in facilitating both the development and maintenance of our adaptive immune system (Opferman and Korsmeyer, 2003; Feig and Peter, 2007; Opferman, 2008). Immune cells undergo a maturation process in the thymus. At that time, those cells that would react against our selves are killed by apoptosis (Medzhitov and Janeway, 2002). A large number of cells are eliminated in the process. Very few T lymphocytes—only about 3%—make it through this rigorous process.

To be effective, the immune system has some special abilities. Most cells divide at a rate that maintains an appropriate population. Others (e.g., neurons and cardiomyocytes) divide much less often. Cells of the immune system are different. They must be ready to multiply rapidly to effectively deal with threats. Equally important, those large populations must return to normal once the threat has passed. In the contraction phase, apoptosis is involved, but the mechanism is not completely clear. Most likely, apoptosis, autophagy, and necrosis all have a role.

Oocytes

Mammalian females (including humans) are born with a large number of eggs (Tilly, 2001). In fact, females are born with several million oogonia. Those are significantly culled before birth, and the loss finally drives menopause when only a few hundred remain. All of those eggs are eliminated by apoptosis. Why are so many eggs created only to be destroyed before birth? The reason is not clear. The process may also involve granulosa cells that provide growth factors, nutrients, and survival factors to the oocyte (Tiwari et al., 2015). The granulosa cells are also subject to apoptosis. Without the support, the oocytes will likely die. One speculation is that it is some sort of quality control or "survival of the fittest" test. Males produce sperm throughout most of their lives. Millions of sperm are released into the lower section of the female reproductive tract. However, those sperm still have a long difficult journey to reach the egg. With eggs, it is different. Each month, a couple of dozen begin to mature, but only one is actually released. The rest are reabsorbed and lost. Perhaps the great loss of eggs is to ensure that only the best is available for fertilization by the strongest sperm. Even after fertilization, many preimplantation embryos are lost, and the reasons are not always clear (Jurisicova and Acton, 2004). There is a balance between pro- and antiapoptotic factors, and perhaps, that balance is disrupted if something is wrong with the embryo.

Red Blood Cells

Erythrocytes or red blood cells are both common and unusual cells. There are lots of them in the body (about 25 trillion!), and they have the critical job of carrying oxygen to cells throughout our body and carbon dioxide back to our lungs. They are unusual because, unlike all other cells, they lack a nucleus or other organelles. They wear out after about 120 days and need to be eliminated from the system. This translates to about 1% of the total pool that must be removed every day. That amounts to about 250 billion cells.

How does it happen? Apoptosis is the most common mechanism for eliminating worn-out cells, but erythrocytes lack nuclei and mitochondria, which are intimately involved in the classic process of apoptosis. Dying erythrocytes do show some features of apoptosis, such as cell shrinkage, membrane blebbing, and exposure of the lipid phosphatidylserine at the cell surface (Lang et al., 2010). To answer this question, some researchers have suggested the term "eryptosis" to describe this pseudo form of programmed cell death (Föller et al., 2008; Pretorius et al., 2016). Eryptosis is a complicated process that involves ion channels and multiple signaling molecules. Many triggers of eryptosis have been suggested. For example, Ca^{+2} ions might cause the cell membrane to form vesicles and the cell membrane to rearrange. Ca^{+2} ions might also stimulate the cysteine endopeptidase calpain, leading to loss of the cytoskeleton and blebbing of the cell membrane. Berg et al. (2001) found that mature erythrocytes contain caspase-3 and -8, but lack other typical mitochondrial apoptotic elements, such as caspase-9, Apaf-1, and cytochrome c. The caspases are active in vitro, not in the erythrocytes. Another protein called calpain initiated morphological changes that resemble apoptosis. Berg et al. suggested that the process in erythrocytes is not exactly apoptosis but it might be a good system for studying aspects of apoptosis. More recently, Dreischer et al. (2022) summarized the current understanding of eryptosis and erythrocyte death. They conclude that it is a stress-induced cell death that differs from hemolysis or senescence, but is similar to another form of programmed cell death called ferroptosis.

Interestingly, erythrocytes in other species of amphibians, reptiles, and birds are nucleated, and they undergo traditional apoptosis (Daugas et al., 2001). So do mammalian and even human nucleated fetal erythrocytes.

Skin Cells

Our skin is the largest organ in our bodies. It protects us from stresses from the environment, such as desiccation, mechanical damage, and infections (Costanzo et al., 2015). The skin comprises three layers (Gilaberte et al., 2016). The deepest layer is the hypodermis. This fatty layer protects the underlying muscles and bones from injuries. Connective tissue holds the skin layer to the muscles and bones, and the fat helps to regulate temperature by insulating the body. The middle layer is the dermis, and it accounts for 90% of the thickness of the skin. Keratinocytes, a type of epidermal cells, are the main cell type. A lot goes on in the dermis. It contains two key proteins. Collagen makes it strong, and elastin allows it to be flexible. Hair follicles, nerve receptors, oil glands, sweat glands, and blood vessels are all found there. The main cells of the inner layers are keratinocytes. The outer layer, or epidermis, contains dead and dying cells and extracellular material (e.g., cholesterol, free fatty acids, and ceramides). The lipids are produced by lamellar bodies and are essential for the waterproof barrier. The pigmentation of the skin is due to melanocytes that produce melanin that is found in the epidermis. It is continually renewed by dividing keratinocytes from the lower layers. By the time the keratinocytes reach the outer layer, they have lost their nucleus, and the cell attachments have weakened. This layer also contains Langerhans cells, which are immune cells that protect against infections.

The skin forms a barrier by various cell-to-cell attachments called tight junctions. Tight junctions form a seal between the epithelial cells, but that seal is dynamically regulated to allow ions, proteins, and even other cells to penetrate it (Brandner et al., 2015). Those tight junctions are facilitated by the shape of the cells themselves. By carefully examining the cells in the ears of mice, Yokouchi et al. (2016) found that the cells resemble a shape called Kelvin's tetrakaidecahedron. In 1887, Lord Kelvin suggested that a 14-sided solid with six rectangular and eight hexagonal sides was the best shape for filling space.

The skin cells protect us, but that protection comes at a cost. The cells are under constant attack from drying, sunlight, abrasions, and insects and other organisms. Estimates vary greatly, but there is no doubt that we lose a great number of these cells every day (Weschler et al., 2011), and those lost cells must be continually replaced.

The replacement process involves a specific type of apoptosis called cornification and the eventual loss of cells by desquamation (Eckhart et al., 2013). Cornification is also involved in the production of other skin structures, such as nails, hair shafts, and the papillae of the tongue. It is a type of apoptosis, but different from typical apoptosis. Unlike apoptosis, cornification results in the death of cells, but those cells are not discarded. They are used as building material to form a protective outer layer of the skin. For this reason, the keratinocytes cannot die too quickly. If that were to happen, inflammatory processes would be initiated that would interfere with the construction of the outer layer. Thus, several protective mechanisms protect the cells from typical apoptosis. First, the cellular organelles are degraded by caspases. Then, cell proteins are cross-linked by transglutamination to form a cell envelope. Finally, the corneocytes are formed into a strongly connected but dead structure. Cornification does not always work, however. The disordered process is generally referred to as ichthyoses, and some forms involve mutations to several genes that function in keratinocyte differentiation and the barrier function of the skin (Schmuth et al., 2013). The result is a thickened layer of outer skin. Treatments address only symptoms and are specific to the particular disorder and patient. No curative therapies are available as yet.

Intestinal Cells

The small intestines perform vital functions (Collins et al., 2022). This long tube (about 7 meters) connects the stomach and the large intestine. It continues the digestion process, and absorbs nutrients and water. In addition, it provides an important barrier that prevents waste and the many microbes in our intestines from crossing the intestinal wall and entering the abdominal cavity. Because the intestines function to absorb nutrients and water, their surface area is folded into villi that protrude from the wall into the lumen. The villi can be seen as tightly packed mountain peaks, and the valleys between them are called the crypts.

One challenge for the intestines is that the continual exposure to digestive enzymes, waste, and microbes results in a lot of wear and tear on the epithelial cells that line the intestines. Over time, those cells wear out and need to be replaced with new cells. So the loss of cells must be balanced by the production of new cells. The key to this system is a cache of stem cells that exist in the crypts and that continually produce new cells. The new cells begin to move up the wall of the villus. At the villus tip, a type of programmed cell death occurs called anoikis (Frisch and Francis, 1994). The epithelial cells of the intestinal wall are anchorage-dependent cells. That means that they sense their position in the intestinal matrix by interacting with the extracellular matrix and the cells near them. When they begin to detach, they detect the loss of those inputs and initiate apoptosis (Gilmore, 2005).

How the cells know to begin anoikis is not clear. Perhaps they simply sense the crowding at the tip. As more cells move up the villus, the tip becomes packed with cells. Some are detached and lost. Another possibility involves inflammation. As apoptosis is initiated, some extracellular vesicles and chemokines may attract immune cells that speed the process. Finally, the microbiota of the intestines may be involved. Certain groups (e.g., *Salmonella*, *Shigella*, pathogenic *Escherichia coli*, *Helicobacter pylori*, and *Cryptosporidium parvum*) enhance apoptosis.

Interestingly, this process may also act as a barrier to some nematode infestations. Nematodes are ubiquitous, and many are parasitic on humans. For example, *Trichuris trichuria* burrows into the intestinal epithelium. They move and feed in the resulting syncitial tunnels. The cytokine interleukin-13 and the chemokine CXCL10 control the rate of cell replacement in the intestine. Cliffe et al. (2005) refer to the actions of those regulatory molecules to speed up the turnover as an "epithelial escalator" that helps to eliminate the parasite.

Cells sometimes lose their ability to sense their proper position, and that loss is a step toward cancer (Weems et al., 2023). As those cells are loosening their attachment to their normal position, they round up and form small protrusions on the cell membrane called blebs. The blebs cause signals that activate prosurvival pathways. If those signals are prevented, the cells continue on their path to death.

Diseases and Aging

Multicellular organisms balance the growth and development of tissues and organisms to ensure their integrity and function. They promote growth with various factors and eliminate damaged or diseased tissues by programmed cell death. Apoptosis is also involved in the pathogenesis of and defenses against diseases (Wong, 2011).

In some diseases, the mechanisms of programmed cell death are circumvented. A reduction in apoptosis or resistance to it can result in cancer (Ouyang et al., 2012). The critical balance between cell proliferation and destruction is lost and cells begin to grow in an unregulated manner so that they infiltrate and destroy normal tissues. Importantly, the balance of pro-apoptotic and antiapoptotic factors is disrupted. The most critical of these factors are the Bcl-2 family (Singh et al., 2019). These mitochondrial proteins promote and inhibit apoptosis. Many blood and solid tumor cancers have increased expression of BCL-2 (Strasser and Vaux, 2020) and also inhibit pro-apoptotic proteins. As a result, the cancer cells continue to proliferate. Because a loss of apoptosis is so linked to cancer, induction of apoptosis is a potential cancer therapeutic strategy. As one example, BH3 mimetics are small molecules that mimic the actions of BH3 and, thus, render tumor cells more sensitive to apoptosis. In addition, caspase function is disturbed (Ghavami et al., 2009). Metabolism and the ability to obtain nutrients also have a role in the progression of cancers and may affect programmed cell death (Sharma et al., 2019).

In some diseases, there is too much programmed cell death. Example include the neurodegenerative diseases, such as Alzheimer's disease, Parkinson's disease, Huntington's disease, and amyotrophic lateral sclerosis (Gibellini and Moro, 2021). In all of these diseases, neurons are lost, and various types of programmed cell death have key roles, including apoptosis, necroptosis, pyroptosis, ferroptosis, and autophagy and necrosis (Moujalled et al., 2021). The causes of the diseases and the cell death are unknown, but cellular stresses and inflammatory processes are involved. For example, amyotrophic lateral sclerosis is characterized by the death of motor neurons in the brain and spinal cord (Moujalled et al., 2021). This leads to muscle weakness and other symptoms. A significant number (20%) of patients have a mutation in superoxide dismutase 1 that causes it to interfere with the activity of BCL-2 and promote apoptosis. Apoptosis is also associated with Parkinson's disease, which features the loss of apoptosis of dopaminergic neurons in the brain. In this case, a mutant form of Parkin cannot ubiquitinate BAK, which stops the action of BCL-2 and thus also promotes apoptosis.

Each of these diseases is associated with the loss of a specific type of neurons, and programmed cell death is also implicated in their pathogenesis. These observations point to potential therapeutic strategies: inhibiting apoptosis or other types of programmed cell death might be beneficial (Erekat, 2022). If so, that would be a great advance since no effective therapies currently exist for these devastating diseases.

Different types of programmed cell death are activated in different infections, and in some cases, the different types work in concert. The genetic programs for apoptosis necroptosis, pyroptosis, and necrosis seem to work together to prevent infections. Pyroptosis may protect against vacuolar or cytosol-invasive bacteria. It is also activated during some infections by HIV (Doitsh et al., 2014).

Importantly, programmed cell death is part of our innate immune system (Jorgensen et al., 2017). Our innate immune system has evolved proteins that recognize molecules from pathogens or others that are released when the cells are damaged (Amarante-Mendes et al., 2018). These pattern recognition receptors activate microbiocidal and inflammatory responses that attack the pathogens. These include immune cell migration, phagocytosis by macrophages and dendritic cells, production of pro-inflammatory cytokines, and activation of apoptosis, necroptosis, and pyroptosis.

Aging is not technically a disease, although some researchers have suggested that it might be. Nevertheless, disorders of programmed cell death are associated with the aging processes (Tower, 2015). For example, autophagy loses effectiveness with aging (Aman et al., 2021). More specifically, the activity of lysosomal proteolysis slows autophagy, exacerbates cellular impairment, and accelerates the diseases of aging. Baar et al. (2017) looked for compounds that could affect FOXO4 and restore apoptosis of senescent cells to slow aging.

Quorum Sensing in Bacteria

Programmed cell death makes sense in multicellular organisms. A problem in one cell (e.g., infection, loss of growth controls) can spread to other cells and potentially put the entire organism at risk. A mechanism that allows the organism to sacrifice some cells to slow infections or during development would be beneficial. Such a mechanism easily fits into our conception of natural selection. But what about single-cell organisms, such as bacteria? Can they have a similar mechanism? Is it reasonable to expect a cell to sacrifice itself to benefit other cells? How does that fit into our theory of evolution?

In terms of sheer numbers, single-cell organisms dominate the Earth. All other organisms have had to deal with the fact that they live in a world dominated by bacteria (McFall-Ngaia et al., 2013). Except for the viruses, bacteria are the most numerous living organisms on Earth. Estimates are that Earth has 5 x 10^{30} bacteria (Whitman et al., 1998) and only 7.5 x 10^9 humans. Bacteria, fungi, and other single-cell organisms live essentially everywhere from the bottom of the ocean to far up in the atmosphere and on every surface. Most are harmless, and many are beneficial. Only a few are pathogenic. Even though there are lots of them, bacteria live in a very dangerous world. They are subject to starvation, drought, temperature extremes, and predation by other organisms. They have protective mechanisms (i.e., toxic proteins that kill other bacteria). Bacteria are all around us.

Most bacteria grow on surfaces and form biofilms (Persat et al., 2015). Bacteria attach to a surface, such as teeth, by using weak molecular forces called van der Waals or hydrophobic forces. Then programmed cell death (Bayles, 2007) causes some bacteria to die, and the survivors use DNA to make a matrix to cement the bacteria together. The polysaccharide matrix encloses the bacteria and protects them from antibiotics and other external forces. Biofilms have been with us forever. Fossils from 3.2 billion years ago have biofilms on them. Biofilms of bacteria include dental plaque and pond scum.

But most critically, they communicate by releasing and sensing signaling molecules (Waters and Bassler, 2005). The process is called "quorum sensing," and they use it to monitor their environment (Allocati et al., 2015) and respond to stressful situations. They signal each other to coordinate gene expression to modulate population density. In that sense, they act like multicellular organisms. They also carry genes for cell suicide just like multicellular organisms. For example, if the density of the colony passes a certain level, some bacteria die. Those signaling molecules have some with the ominous-sounding name of extracellular death factors. They differ among species. One in *Pseudomonas aeruginosa* results in large amounts of reactive oxygen species that lyse the cells (Hazan et al., 2016). In other organisms, programmed cell death involves a balance of toxins and antitoxins (Buts et al., 2005) that the bacteria use to promote or hinder cell growth. The balance between the production and degradation of the two compounds allows the cell to control its growth.

No matter which system they use, the bacteria regulate their population by having some bacteria initiate a sequence that kills the cells. Even though individual bacteria are lost, most survive by using the nutrients supplied from their dead comrades until conditions improve.

So bacteria use quorum sensing and programmed cell death to control the population of the colony. Yet, bacteria are single-cell organisms. How can they benefit from this mechanism? It seems to violate the concept of natural selection for one bacterium to sacrifice itself to save others. It's also difficult to imagine an evolutionary pressure that would select for such a mechanism. Nevertheless, quorum sensing is a fact. The explanation must lie in the idea that the colony behaves as if it were a multicellular organism. The selection for this mechanism must be that it promotes the survival of the most cells.

Cell Death Supports Cell Life

The most amazing aspect of programmed cell death is that it allows cells and organisms to use death to facilitate development and protection against disease. Life has learned to use death to support life. We might not be able to see the antibody-secreting cells that do not attack our own tissues, but we are certainly grateful that they are eliminated by programmed cell death. Nevertheless, we can see the results of these processes in our daily lives. We have fingers because the excess tissue between them is eliminated by these death programs. Ripe fruit and leaves in the autumn fall because a layer of cells is programmed to die at a specific time. Metamorphosis causes caterpillars to transform into butterflies and tadpoles into frogs. Even bacteria use programmed cell death to reduce populations so that some can survive. Programmed cell death is one of the most amazing processes in biology.

12

Laughing and Crying

Crying is our first communication with the world. Babies cry upon birth. Who wouldn't? One moment we are in a warm, protected environment, and then next, we are shoved out into the cold world surrounded by strangers and strange sounds. In the classic image, we are dangled upside down and smacked on the rump. We can't really see, and we don't speak the language. A strong protest is definitely in order.

The smack on the rump is no longer used. Nurses or other healthcare providers rub the baby gently with a warm towel to get them to cry. But the actual crying is important. It means that the baby is breathing. They have successfully made the transition from obtaining needed oxygen from the mother via the placenta to breathing the air around them. They also need to expel any amniotic fluid or mucus from their lungs and trachea. Once they cry, their own lungs are working properly and the airway is clear. The baby and everyone else can quite literally breathe easy.

Crying is part emotional and part physical. From the physical perspective, crying is only one of our emotions that register on our faces. Human faces have 42 muscles that allow us to produce myriad facial expressions that can signal many different emotions. Early on, facial expressions and vocalizations are all we have to express ourselves. Later, we use those muscles to laugh and cry and register surprise, anger, fear, and more. With that amazingly complex repertoire, we can greet friends and strangers, show our pleasure or pain, or warn others. We can communicate deep emotions even when words fail us. But how did it come to be that our emotions and physiology combine in such prominent displays?

Crying and Tears

Crying often involves tears, but not all crying is associated with tears. Human babies cry from birth, but they don't produce actual tears for the first 2–3 months (van Haeringen, 2001). In fact, tears are not always associated

The Biology of Us. Gary C. Howard, Oxford University Press. © Oxford University Press 2024.
DOI: 10.1093/oso/9780197664797.003.0012

with crying. Tears are important even when we are not crying. They lubricate our eyes. We blink 10–20 times a minute, and every time we blink, our eye lids spread a thin layer of tears across the surface of our eyes that cleans the eyes while wiping away any dust or other material. Tears are secreted by tear glands that are just above the eyes. They drain down through the tear ducts and exit at the inner corners of our eyes.

Tears taste salty, but they are more complex than just salty water. In fact, tears form three layers on the surface of the eye, and each layer has its own distinct source, composition, and important function. The outer most layer of the tear surface is the lipid layer. It is secreted by the meibomian glands that are found at the margins of the eyelids near the eyelashes (Abelson et al., 2016). About 20–30 glands on the lower lid and 40–50 on the upper lid secrete a mixture of fats and oils. Those lipids reduce evaporation, seal the lids during sleep, and help to keep the eye's optical surface nearly perfect. The middle layer is a complex aqueous solution of many components, including electrolytes (sodium, potassium), urea, amino acids, carbohydrates, nucleotides and nucleosides, peptides, phospholipids, antibodies, lipocalin, lactoferrin, lysozyme, lacritin, and various other biochemicals. Different species of animals show different compositions of tears (Raposo et al., 2020). The differences likely reflect the different needs resulting from different environments. The innermost layer is the mucin layer. The cornea and conjunctiva produce mucins that span the membrane, and the conjunctiva produces soluble mucins (Hodges and Dartt, 2013). Mucins are proteins that are rich in the amino acids serine and threonine and are linked to various oligosaccharide (sugar) side chains. They hydrate the surface of the eye, lubricate the eye and eyelid during blinking, and hinder the binding of pathogens to the eye.

To make tears even more complex, the composition varies with the cause of the tears. Basal tears are the normal tears that are spread across the eyes by blinking. Strong irritants (e.g., onions) can cause reflex tears. These extra tears flood the eyes in an attempt to wash away the irritant and protect the eyes. Finally, strong emotions, sad or happy, can cause tears. These emotional or psychic tears have a different composition. They contain more hormones, such as prolactin, adrenocorticotropic hormone, and leu-enkephalin, than the other two types.

The solid components of tears are familiar to all of us. We wake up with deposits of material, especially in the inner corners of our eyes. We call it "sleep," but its proper name is rheum. While we are sleeping, we do not blink, and these tear components dry up. Sleep or rheum is normal, and we ordinarily just wash it away. However, discharges from the eyes can also signal trouble. Sticky or goopy eyes might indicate an infection. Conjunctivitis or pink eye is a common infection in small children. Tear ducts can also become blocked, resulting a condition called dry eye.

So tears make great biological sense in maintaining the health of the eye. But how did they come to be associated with our emotions? Shedding tears seems to be a very human characteristic. No animal is known to cry. Why do we? Some scientists speculate that crying might have evolved as a form of intimate communication. A crying person is usually in a vulnerable state. Yet, tears fall silently. Only those close to the person crying will notice. Thus, the crying person can communicate their despair very quietly to another without alerting an enemy. Furthermore, the tears do more than communicate. They can elicit empathy in those observing the tears.

Babies cry a good bit, but why? Lummaa et al. (1998) suggest four reasons for infants to cry. The reasons are not mutually exclusive. First, the infant cries to regain physical contact with the mother. Second, crying communicates information on fitness that attracts more attention to the infant. While these actions might be perceived as unsettling, there is evidence for it. For example, Worchel and Allen (1997) asked 40 first-time mothers to listen to recordings of babies crying and to use those sounds to determine the state of the baby. The babies were full-term and low-birthweight premature (LBWP) newborns and 6-month-old full-term and LBWP infants. Interestingly, the mothers reacted less positively to the cries of the LBWP infants than those of full-term infants, and the mothers of LBWP infants tended to withdraw from a premature infant's cry rather than respond to it. Third, crying "blackmails" parents into doing that the infant wants to keep it quiet in the face of possible predation. Fourth, crying tends to show strength, and so parents may delay adding a sibling to the benefit of the crying child.

Of course, it isn't just babies who cry. Emotional crying is universal in adult humans (Gračanin et al., 2018). Over time, crying changes. At 9–11 months, children become afraid of strangers and strange places. At 3–4 years, children can signal others with words and other means, and crying becomes less intense. Humans have facial muscles that allow many expressions, and the face is a natural place for another to look for signs of a person's emotional state. Tears are another layer of expression that allows us to show sadness and suffering. Some scientists attribute tears and weeping to the need to signal those emotions. They could also be a signal of submission during fighting.

Of course, people also have tears when they are laughing, but those are more likely the mechanical result of the contraction of eye muscles that control the expression of tears.

Do animals cry too? This concept is somewhat controversial (Lingle et al., 2012). There are multiple anecdotal examples of weeping animals, but hard evidence is more difficult to come by. Infant animals vocalize when they are trapped by humans or separated from their mothers. Is that comparable to human crying? Some regard those vocalizations as simply involuntary responses rather than related to emotions. Other scientists point to differences in strategies for avoiding predators. For example, isolation calls are greater from young that travel with their parents and fledglings that can more easily be separated from their parents vocalize more than those who are safe in a nest. Physical stresses (e.g., hunger, cold) also elicit vocalizations.

Laughing

Laughing seems so simple and natural, but there is more to it than meets the eye. And it's the eyes that are one of the primary determining factors. In 1862, the French anatomist Duchenne de Boulogne noted a difference between a "genuine" smile that comes from true emotions and a "fake" smile that we consciously assume. All smiles involve the zygomatic major muscles that lift the corners of the lips. It's easy to activate those muscles and smile. A genuine or Duchenne smile (as it is now known) results from our true emotions. It also involves the muscles around the eyes, the orbicularis oculi. Those muscles cause the skin around the eyes to wrinkle up to form crows' feet.

Paul Ekman at the University of California, San Francisco, has made seminal contributions to the study of smiles. He showed that, unlike other smiles, Duchenne smiles are associated with activity in the left frontal cortex, an area of the brain also associated with enjoyment. They described five indicators that differentiate types of smiles (Ekman and Friesen, 1982): (1) muscles in the eyelids (orbicularis oculi) and those that connect the cheekbone to the corner of the mouth (zygomaticus major) are activated together (Duchenne's smile), (2) the zygomaticus major contracts on both sides, (3) the actions of the zygomaticus major are smooth and regular, (4) the length of the action of the zygomaticus is consistent from one smile to another, and (5) the zygomaticus major and orbicularis oculi reach maximal contraction simultaneously. Ekman and colleagues reported a series of experiments that showed that genuine pleasure-filled smiles can be differentiated from fakes (Ekman et al., 1990).

However, not everyone is convinced that smiles are so definitive, and in more recent years, some biologists have suggested that even Duchenne smiles might be faked. Krumhuber and Manstead (2009) tested volunteers to determine whether they could differentiate between a spontaneous Duchenne smile and a posed smile. Observers could pick out real pleasure smiles from posed smiles only under limited circumstances. Thus, the researchers were forced to doubt the reliability of Duchenne smiles and to wonder if Duchenne smiles can be faked.

Facial recognition has become commonplace, and not surprisingly, efforts have also been made to extend those capabilities to determine emotional states (i.e., automatic emotion recognition). Kaur et al. (2022) tested these systems against the self-reports of participants. They found that the programs performed poorly in predicting emotional states and suggest that more work is needed before these programs can replace actual interactions with individuals.

Laughter is a universal human characteristic (Gervais and Wilson, 2005). It begins at 2–6 months of age, although babies might smile before that, and seems to be genetically programmed into us. Infants spontaneously laugh when presented with novel stimuli in a nonthreatening manner. "Peak-a-boo" games with their mothers are a great example. However, those same gestures might elicit crying if presented in a more threatening manner. When playing, primates and particularly the great apes exhibit a "relaxed" face and often a sort of panting sound that some have speculated is equivalent to human laughter. That suggests that our ancestors were laughing before the primate lineages split from our human line about 6.5 million years ago. In addition, Ross et al. (2009) examined the acoustics of vocalizations in various apes (e.g., orangutans, gorillas, chimpanzees, and bonobos) and humans that were induced by tickling. They found that the relationships generated from the acoustic data recapitulated those from comparative genetics.

The origins of formal humor, the type of humor based on language, is more controversial (Sabato, 2019). Plato and the Greek philosophers thought that we laughed at earlier versions of ourselves and the misfortunes of others. Freud thought it was a nervous outlet that involves scatological and sexual themes. Still another theory posits that we laugh at incongruous situations.

Like smiles, laughter comes in two types, and like smiles again, they are named for the French scholar Duchenne. Duchenne laughter is the spontaneous, genuine response to something funny. Non-Duchenne laughter is an intentional response to a social situation. They involve different nerve connections and different evolutionary origins. Duchenne laughter likely originated as a method to indicate a safe environment and to promote social bonding. Non-Duchenne laughter reflects a more calculated origin that may betray a hint of dishonesty.

Anatomy of Speech

We use speech to let others know our social status, our personality, and our emotional state. It is an inherent part of our being. Loss of this ability (e.g., sequela to a stroke) is traumatic to the patient. Our ability to communicate requires certain anatomical structures. Speech, in particular, involves the interaction of the fluid air with the fleshy folds in the larynx in our throat (Zhang, 2016). The cries from babies and the words spoken by each of us every day require certain physical structures (Lieberman, 2007). The process begins with the air flow from the lungs through the larynx and is further refined by the mouth and nose. Each part of the anatomy has a role.

The vocalizations in most tetrapods (including humans) result from the anatomy and physiology of the tissue structures. Thus, many of our human vocalizations derive from common biological evolution with other animals. For example, vocalizations are generally produced in two steps (Matzinger and Fitch, 2021). The air that powers vocalizations typically comes from exhalations in most tetrapods. However, some animals (e.g., donkeys, chimps) can also vocalize while inhaling. In step 1, air from the lungs is forced through the larynx in most tetrapods and the syrinx in birds. These tissues create sounds at particular frequencies. In step 2, those sounds are further processed in the supralaryngeal vocal tract to enhance or attenuate specific frequencies. Beyond those two steps, humans have continued to refine vocalizations cultural and linguistic influences.

Speaking

From their very first cries just after birth, babies are communicating. Every baby is different and develops at its own rate. However, there are some general milestones for how they begin to talk. For the first 3 months, they smile and coo. Of course, they also cry when they are unhappy. By 6 months, they begin babbling. Sounds, such as "puh," "buh," and "mi" are common. They also laugh, giggle, and make other sounds. At around 12 months, they put their sounds together to form strings of the same sounds (e.g., "ba-ba-ba-ba"). At 12–18 months, they may begin using single words, and those might be ma-ma, da-da, or cat. By 18 months, they will probably have about 50 words, and by their third birthday, most children are speaking in full sentences. Babies learn languages with incredible ease, whatever the language they are immersed in (Kuhl, 2004). Interestingly, Lameira and Moran (2023) speculate that "p" (as in Pablo Picasso) is perhaps the oldest vocalization and preceded all spoken languages. It is found in every known language. It is the first sound made by human infants. Finally, most primates can also use their lips to make a raspberry sound that involves the sound of "p." It's an interesting observation and potentially a persuasive case.

Here is a quick overview of the process of speaking. The larynx (or "voice box") is a cartilaginous structure in the throat that holds the vocal cords. The vocal cords are two small flaps of muscle. Each is about a centimeter long. Folds in the cords are about 11–15 mm in adult women and 17–21 mm in men. Vibrations of the cords create speech. The glottis is the space between the vocal cords. At the back of the throat, the pharynx is a tube that connects the larynx with the mouth. Several parts of the mouth are involved in speech. The palate refers to the roof of the mouth. The hard palate is in the front and the soft palate or velum at the back is not supported by cartilage. The uvula is a small tissue that dangles at the end of the velum. It plays a role in the pronunciation of some consonants. About a centimeter behind the upper teeth in the hard palate is a bony structure called the alveolar ridge. The tongue is very important. This muscle changes shape to aid in pronunciation. Even the teeth are used to make some sounds. Finally, the lips help to form the sounds of consonants (e.g., p, b, m, f, v, and w) and vowels. We use our tongue, lips, and teeth to speak. The tongue is a group of eight muscles, and like any muscle, they require strength and control to function properly. The eight muscles are divided into two groups. The four extrinsic muscles are attached to bone and change the tongue's position. The four intrinsic muscles are not attached to bone and allow us to change the shape of the tongue. A fibrous tissue, called the lingual septum, runs from front to back at the center of the tongue and forms a slight grove called the median sulcus. The actions of the tongue in speaking are more or less the opposite of those of swallowing. In speaking, the tongue tends to move up and forward, and in eating it moves down and backward. It can produce more than 90 words per minute. The tongue is important in pronunciation. It is key to being able to pronounce certain letters and especially the consonants. For example, the tongue enables us to pronounce "t," "d," and "l" and to roll the "r." For the letter "k," we make the tongue slightly more narrow at the rear. For the letter "s," we pull the tip of the tongue backward. Even the teeth are used to make some sounds. Finally, the lips help to form the sounds of consonants (e.g., p, b, m, f, v, and w) and vowels. Some people have trouble moving their tongues appropriately. In some, the tip of the tongue stays between the teeth. That action results in a lisp. Parrots are well-known for imitating human speech, and their tongue is important to their mimicry. However, their anatomy is much simpler than that of humans and only allows them limited speech. Their tongues are very thick, and that helps them to make the sounds of human words. They use their tongue to form the words.

The other organ that is intimately involved in speech is the brain. The brain has many parts, and the part associated with speech is called Broca's area, which is located just above and behind the left eye in the left hemisphere of the cerebrum. This area has been recognized as important for speech for over 150 years, but only recently has its specific function been understood. Flinker et al. (2015) at Johns Hopkins carefully mapped the electrical activity during speech at the cortex of patients undergoing treatment for epilepsy. Broca's area is most active just before we speak words. It seems to develop the plan for what will be said and then to monitor the words and make corrects as needed. Its role may be between the organization of sensory inputs by the temporal cortex and the actions of the motor cortex that directs the physical actions of speech. Other studies have continued to refine the brain areas involved in speech. Gajardo-Vidal et al. (2021) found that brain areas adjacent to Broca's area are also involved in speech and damage to Broca's area is not necessarily injurious for speech. Speech processing in the brain continues to be an area of intense research. One area that is very exciting focuses on detecting brain activity so that intended speech can be translated into speech artificially. Various methods are being developed to help those with speech difficulties. Some use facial recognition programs to translate facial expressions into speech, but these suffer from movements, changes in angles, and lighting. Others are designed to detect electrical activity in various areas of the brain (Proix et al., 2022) or use sensors placed on the face to detect and translate muscle movements into speech (Kim et al.,

2022). Many of these sound like science fiction, but it is an exciting area of research with great potential.

Hearing

Communication involves both the transmission and reception of information. In terms of spoken language, we also need to hear what has been said. The great biologist E.O. Wilson (1975) described communication as information transmitted from one organism to another that allows the receiver to act on that information. Wilson understood communications well. He focused his career on the study of ants and other social insects. Those insects live in large colonies and must communicate effectively for the colony to survive. Wilson extrapolated from his insect studies to develop the field of sociobiology.

We hear when sound waves enter our outer ear and travel down the ear canal to the eardrum (Schwander et al., 2010). The eardrum vibrates according to the sound waves, and those vibrations impact three tiny bones (i.e., the malleus, incus, and stapes) in the middle ear. Those bones amplify the vibrations and pass them on to the cochlea in the inner ear. The cochlea is filled with fluid. The elastic basilar membrane divides the cochlea into upper and lower parts. On top of the membrane are hair cells. As the sound vibrations enter the cochlea, they cause movement of the fluid that, in turn, moves the hair cells. Hair cells near the end respond to higher-pitched sounds, and those further along respond to lower-pitched sounds. As the hair cells move, they hit the top of the cochlea, and that impact results in an electrical signal that is carried by the auditory nerve to the brain. Interestingly, signals go to the auditory cortex on the opposite side of the temporary lobe. In other words, signals from the right ear go to the left side of the brain, and those from the left ear go to the right side. There the signals are processed and interpreted.

Vocalizations and Language

The ability to communicate effectively is critical to animals (Gillam, 2011). They use vocalizations and other modes of communication to attract mates, warn of predators, protect territory, and more. Animals use many mechanisms to communicate. Animals that are active during daylight hours often use visual means of communications. For example, the bright colors of some birds can be easily seen. Fireflies (family: Lampyridae) use bioillumination so they can find mates at dusk. Other animals use sounds and vocalizations. Cicadas (family: Cicadidae) sing to attract mates. They produce their song by using their abdominal muscles to buckle a structure on their sides called a tymbal. Their songs can be very loud (up to 120 decibels). Animals also use chemical signals to attract mates. The chemicals are powerful but diffuse easily. Social animals have more sophisticated means of communications that reflect their cooperative society and need to share responsibilities. Subordinate animals display submissive postures and sounds, and dominant animals indicate their peaceful intent. Finally, communications can sometimes backfire. Some species have learned to listen for calls that direct them to their prey.

Communications among the great apes are quite complex. They use extensive vocalizations, but their repertoire of gestures is at least as large. Hobaiter and Byrne (2011) observed 66 distinct gestures among chimpanzees in the wild. They note that 24 gestures are found across species, including gorillas, orangutans, and chimpanzees. These numbers are conservative, since the researchers put strict limits on how they defined a gesture. Most importantly, it had to be intentional, and the sender had to look for an appropriate response. They were left with a question of why the chimps and other apes have not expanded their set of gestures.

While there is some disagreement among experts on what constitutes a language, there is general agreement that nearly 7,000 languages are spoken in the world, and new ones are being discovered every year (Erard, 2009). Indeed, formal language is one characteristic that still seems to be unique to humans. How and when did humans learn to speak? That question is further compounded by the fact that written language has existed only for the last 6,000 years. Before that time, we must rely on studies of human anatomy. When did the physical elements needed for speech appear in the fossil record?

Some scientists suggest that, about 100,000 years ago, language began when changes evolved in the human mouth and pharynx and the brain increased in size. The structural changes allowed humans to have more control over the vocalizations. Others believe that onomatopoeia was important in the development of human language. Early humans imitated sounds that they heard in the environment or vocalizations that communicated joy, pain, or surprise. It is easy to imagine that the early sounds might have been warnings about predators or other dangers. Still others believe that humans began communicating with hands that we freed by bipedal locomotion. Later those hand gestures evolved into spoken language. Observations of modern sign languages support this theory. In fact, the language areas of the brain are also used for control of the hands and arms. The tools used by early hominids also advanced very slowly for some time. Could the use of the hands for communications have slowed tool development? Spoken language is thought to have been the main form of communication among humans by about 50,000 years ago. Other prominent theories of language development emphasize the importance of social interactions. Humans needed to share information about hunting, defense, and other matters, and language served this purpose. As they developed more complex lives, they may have needed more sophisticated means of communicating. Of course, these theories are not mutually exclusive. They might all have played a role. In addition, we do not know how many times language evolved. Was it a single event or did it occur in multiple places and times? And how different might those situations have been?

Most studies of the origins of spoken language focus on the differences between the abilities, cognitive and functional, between humans and our close relatives, the great apes. What do we have that they do not or vice versa, and could that explain our mastery of language? In recent years, we have come to more fully appreciate communications among animals, and some few primates have been taught American sign language. None has yet come close to the vocabulary of about 1,000 words of a typical 3-year-old human, but they clearly have the capacity to learn a language. We and they also seem to have fairly similar anatomies and the ability to make sounds.

In a recent paper, Nishimura et al. (2022) found an interesting difference in humans. They used modern imaging technologies to examine the larynges from 29 genera and 44 species of primates, including humans. They found a membrane in the primates that humans lack. Then using mathematical modeling, they examined the effect of that extra membrane on the vibrations produced by vocalizations. Without the membrane, human sounds are more stable and conducive to language than the far more complex sounds that are important to primate communications.

A separate issue concerning language is counting. Studies of primitive tribes show that they have very limited understanding of and few words for counting (Gelman and Gallistel, 2004). Those might include one and two, but three and four and higher numbers are rarely or never used. There seems to be a system that allows humans, infants, and nonhuman animals to process numerical information at least to a limited extent. All seem to understand distances and reaction times on some level, but the relationship of those capabilities and abstract thinking is not clear. Cantlon et al. (2006) supported this hypothesis by using functional magnetic resonance imaging to test adults and children on their ability to respond to different numbers and shapes of arrays of symbols. They found that the children had a similar response, even though they lacked ability with numbers. This is a fascinating area of research.

Evolution of Speech

Although animals communicate, they do not have the complex spoken and written languages that humans have developed. Even our closest relatives cannot speak. A small number of great apes (e.g., the chimpanzees Viki, Washoe, and Nim; the orangutan Chantek; the bonobo Kanzi; and the gorilla Koko) mastered an impressive number of signs so that they could communicate with researchers (Wayman, 2011). Like their wild relatives, they all could vocalize, but those sounds lack the complexity of human language, and none could speak. They lack the anatomical features that provide humans with the physical basis for language to develop (Lieberman, 2007).

How and when in our evolution did those abilities appear? By about 70 million years ago, mammals had developed a number of the anatomical structures that would enable later human speech (Fitch, 2018). The larynx of mammals has a thyroid cartilage and the thyroarytenoid muscle is found in the vocal folds. They also have a fleshy tongue, lips, and a soft palate and the nerve connections to use this anatomy. These developments allow mammals greater control over their vocalizations. The evolution of our anatomy continued in early hominids (for an extensive discussion of these changes, see Lieberman, 2007). The face became flatter and the mouth shortened, compared to apes. The tongue descended into the pharynx to become more rounded, and the neck became longer. These structures appeared relatively late (about 50,000 years ago) in human evolution. The ability of our close relatives, the Neanderthals, to speak is the subject of some debate.

Some of the most intriguing findings come from genomic studies (Liu et al., 2021). Katherine Pollard and colleagues discovered sequences in the human genome that seem to have evolved much faster than those of our close relatives (Pollard, Salama, King, et al., 2006; Pollard, Salama, Lambert, et al., 2006; Hubisz and Pollard, 2014). To identify those sequences, Pollard compared the total genomes of chimps and humans with sophisticated statistical analysis and found specific sequences that experienced multiple changes since humans diverged from chimpanzees millions of years ago. She termed these sequences "human accelerated regions" or HARs.

The HARs are fascinating for several reasons. First, they are not located in the regions of the human genome that encode proteins. Those regions include only about 2% of the total genome, a surprisingly small fraction. They are found in the so-called dark genome. The dark genome comprises the great majority of the human genome. Research has identified the functions of some sequences there. For example, some encode RNAs that regulate the expression of genes in the encoding genome. But the majority have no known functions.

Second, the HARs are not randomly distributed in the genome. They tend to be positioned with other HARS near developmental transcription factors that control the activity of other encoding genes and gene pathways. Several are known to serve as genetic enhancers of other genes. Tissues that show more expression of these sequences than others include limb, eye, forebrain, and the midbrain-hindbrain boundary. The intriguing aspect is that some of these regions seem to be involved in characteristics that differentiate humans from the great apes, such as an opposable thumb and spoken language.

Several theories have tried to account for the evolution of human speech and language. Many animals, including the great apes, vocalize to communicate. One of the most interesting involves gestures. Language, in its broadest sense, is not limited to speech: Gestures with the hands, face, and rest of the body are part of our ability to communicate. Most humans gesture while talking, and the American Sign Language is an entire language based on hand movements. Many of us use hand gestures routinely with or without spoken words. For example, the Romans used a thumbs up or down to determine the fate of fallen gladiators. We point at things. We applaud performances that please us. Many cultures have gestures that signify displeasure or are even quite rude.

Like humans, many animals have extensive repertoires of vocalizations that they supplement with gestures. Their communications come close to the size of our vocabulary, but gestures are an important component of their communications. We can assume that early hominids also used gestures to communicate even before the advent of spoken languages. In fact, Arbib et al. (2008) argue that the use of gestures is one of the early evolutionary characteristics of protolanguage and is critical to the development of spoken language. Their "mirror system hypothesis" is based on the location of the control of gestures in Broca's region of the brain that has long been associated with speech. They believe that gestures predated spoken language, but served as a springboard to that spoken language. Animals lack "compositionality," which refers to the ability to combine words into phrases.

When did spoken language begin? That is very difficult to determine. Clearly, there is no record as there is with written language. In addition, most animals communicate. Most likely, there was a continuum from the vocalizations and gestures of our hominid ancestors to modern spoken language. That development depended on the appearance of the anatomy and genetics that would support speech. Lieberman (2007) estimated the correct anatomy was in place at least 50,000 years ago. Others took a linguistic approach. Perreault and Matthew (2012) examined the length of time needed to realize modern phonemic diversity and concluded that language began 60–70,000 years ago, an estimate that seems to be consistent with that of Perreault and Matthew.

Although some individual animals have learned some words of a human language, they still struggle to be able to truly use those words effectively. Even when animals, such as birds, have a considerable vocabulary of sounds, they still cannot "say" many things. Beecher (2021) also raises a serious question: In the case of animal communication, are the sender and receiver in agreement on that communication or are they simply trying to exert control? He argues that our ancestors lived in a more communal existence and, thus, had to cooperate to a greater extent. Other species lack this social environment. Some animals engage cooperatively. Beecher mentions honeybees, which use a dance to transmit high-quality information to their colleagues (e.g., location of honey or an attractive hive location). This cooperation confirms his general thesis on the importance of the social environment of early humans.

Searcy and Nowicki (2005) pose an interesting question about animal communications: Is the information that is sent accurate and reliable for the receiver? Is the sender sending reliable messages or not? It could be a simple question. Humans have a lot of problems with integrity. Are they alone or are animals also involved? However, it's a much more profound question than it might seem at first glance. If selection acts at the group level, then one would expect that animal communications would be honest. Yet, we know that selection works at the individual level. In that sense, male animals might make themselves look better than they are to attract a mate. Other animals might make puff themselves up to look bigger and stronger than they arc to avoid a predator. Still, the receiver must receive reliable information. In the end, the interests of the sender and receive both need to be satisfied for the system to work.

We can gather a lot of information about a person even if we don't understand their words. Latinus and Belin (2011) give an excellent example of this. They ask us to imagine a couple speaking a foreign language and sitting behind us on an airplane. Although we can't see them, we can know a lot about them just by listening. We can imagine their gender, age, mood, and more. Males tend to be larger than females and children, and vocal frequency is a function of the size of the vocal folds. Other characteristics of the sounds can convey information about the relationship of the individuals and their moods.

Conversations between humans are characterized by a regular pattern of taking turns (Sacks et al., 1974). Each person speaks in turn, more or less, for a brief period. This arrangement allows two or more speakers to participate in a conversation alternating so that each has a chance. The form of each brief communication includes clues to when that speaker will complete their thought so that others can have a turn. The gaps between speakers are very short. Interestingly, the interactions do not follow any artificial "rules." They are locally governed by the two or more actors (Wilson and Wilson, 2005). Of course, this does not always work. Some people hog a conversation, and others talk over each other. However, interruptions are looked on as rude. Thus, it is interesting that it works as well and as completely as it does, regardless of the language or culture. Although turn-taking is a very human characteristic, some researchers have hypothesized that it also occurs in animals (Pika et al., 2018).

Chemical Communication

The baby's cries upon birth are not the first communications from the fetus to the mother. Labor begins after some trigger initiates the process of parturition. The process is amazingly complex and involves multiple hormones, proteins, and biological clocks (Menon et al., 2016). That trigger is not clear.

Some theories have implicated production of sufficient amounts of the surfactant protein SP-A (Condon et al., 2004). Surfactant allows the young lungs to function appropriately once the infant is in the open air afterbirth. It contains proteins and lipids with both hydrophilic and hydrophobic regions, and they function by reducing the surface tension at the interface of the air and water in the alveoli in the lungs. Others have disputed this idea, but clearly, some signals pass between the mother and child to initiate childbirth. SP-A has a role in parturition (Depicolzuane et al., 2022), but it is not likely the only signal passed from the fetus to the mother. Experiments in mice showed that loss of SP-A only delayed birth by a number of hours. That observation suggests that regulation of the process is more complex and that additional factors may be involved. Mendelson et al. (2019) showed that multiple factors have roles.

A quick web search will turn up dozens of advertisements for human pheromones meant to attract the opposite or same sex with many promises and guarantees to make someone sexually irresistible. Many have names that combine semiscientific terms with words with clear sexual connotations. Pheromones have been identified for many animal species. Why not humans? Could all of those dozens of products be wrong? In fact, human pheromones might exist, but after nearly 50 years of study, none has been found despite numerous claims (Wyatt, 2015).

Pheromones are well-documented chemical signals that are used by some specific animal groups for communications among members of that species. Those signals elicit behavioral or developmental responses or both. Pheromones are best studied in insects (Vickers, 2017). Insects use them to attract or find mates and more. As one of the earliest observations of pheromones, Jean-Henri Fabré captured and caged a female giant peacock moth (*Saturnia pyri*) at his home (Fabré, 1914, pp. 179–180). His son later called him to see the many males that had been attracted to the female. The female was secreting pheromones to find a mate, and she was very successful. Ants release a trail of pheromones so they can find their way back to the nest. Bees use them to alarm the hive about an attack. Cats and dogs use them to mark their territory.

In 1959, Adolf Butenandt was the first to chemically characterize a pheromone (Butenandt et al., 1961). Bombykol is secreted by female silk moths (*Bombyx mandarina*) to attract males. The males flutter their wings in response, and Butenandt used that response as a bioassay so that he could isolate the molecule responsible for the action. Bombykol is now used to protect crops. Minute quantities sprayed on fields confuse the males so that they fail to find a mate.

The social insects (i.e., ants, bees) use pheromones extensively. First, the queen releases a specific pheromone. This molecule causes other insects to develop as workers since there is already a queen. For example, queen bee larvae are fed only royal jelly, which is a special food produced by nurse worker bees. The exact component of royal jelly that determines the queen is controversial (Maleszka, 2018), but whatever it is, it has a profound effect on the queen. She is far bigger than the workers and she lives far longer to lay nearly a million eggs.

Mammals (e.g., elephants, goats, and pigs) also have pheromones, but the mouse is most studied. Mice are nocturnal animals, and the males leave trails of urine drops that contain a pheromone darcin. Females remember where the darcin and other identifying smells are found and return to that spot to mate with the male who "owns" that territory.

Like other smells, pheromones are detected by sensors (DeMaria and Ngai, 2010). Silk moths have sensors in their antennae. Mice have 10,000 olfactory sensory neurons (Xu et al., 2020) in the roof of their noses that work in various combinations to identify smells. Chemoreceptors on the surfaces of these cells are proteins that bind the smell molecules. Each receptor binds only to a specific smell molecule. The binding causes a signal to be sent from the cell to the olfactory center in the brain.

The search for a human pheromone has focused on four molecules: androstenone, androstanol, androstadienone, and estratetraenol. Those searches have, for the most part, focused on axillary sweat, which contains those molecules. One problem is that humans lack a vomeronasal organ. Lizards, snakes, and some mammals (e.g., cats, dogs, cattle, pigs, some primates) have it. It is located in their nostrils and used for detecting pheromones and other chemical signals.

Still, there may well be a human pheromone. It seems likely that they exist. But in spite of the myriad advertisements for a magic elixir for attracting the opposite sex, no human pheromone has yet been identified.

Conclusions

We humans use communications at every level. During development and throughout our lives, we depend on the cellular communications (e.g., hormones, neurotransmitters, transcription factors, microRNAs) that allow us to grow and thrive. Many human expressions that show emotions resemble reflexive behaviors that protect the body, such as the startle reflex and blocking and withdrawing reflexes (Graziano, 2022). The reactions are very fast. They can take only a fraction of a second. Graziano (2022) wondered if they might be related and if that might provide a basis for the evolution of emotional expressions. Other physical means of communication include facial coloration (e.g., blushing, red-in-the-face, white as a ghost) (Thorstenson et al., 2021), facial expressions, teeth grinding, and overall demeanor. As we grow and develop, we gain the ability to better control those expressions and emotions.

13

Sleep and Consciousness

Sleep is the most idiotic of all behaviours.... Indeed, it has been said
that if sleep does not serve an absolutely vital function, then it is the
biggest mistake the evolutionary process has ever made.

—Matthew P. Walker (2021)

We all need sleep, and we feel great after a good night's rest, but its purpose
is still not completely understood. As far as is known, all animals require
sleep. Sleep is critical to our health and well-being, and yet, when we sleep,
we are not eating or reproducing. Worse still, we and all sleeping animals are
highly vulnerable to predation while sleeping. We are relatively defenseless.
The evolutionary benefit of sleep must be enormous to overcome the seemingly negative pressures against sleep. But why would evolution put us in this
position?

Was it "the biggest mistake the evolutionary process has ever made" as
Professor Walker (2021) asks? His answer to the question is a resounding
"No," and he makes a solid case. Every physiological system in our body
needs sleep. Blood sugar levels, the immune system, antibody responses,
hormonal levels, and more are affected by a lack of sleep. The brain is particularly vulnerable to a loss of sleep. In addition, several disorders affect the
sleep of many people.

We aren't alone in our need of sleep. All animals that have been studied
seem to have some form of sleep, including invertebrates. Plants might even
need some form of sleep or rest. From those observations, sleep seems to have
developed early in evolution. That level of conservation indicates a powerful
evolutionary pressure to maintain sleep despite its seeming shortcomings.
The benefit gained from the repair and restoration of our physiological
processes greatly outweighs the disadvantages of being more vulnerable to
predators when sleeping.

The Biology of Us. Gary C. Howard, Oxford University Press. © Oxford University Press 2024.
DOI: 10.1093/oso/9780197664797.003.0013

Yet, perhaps the most fascinating aspect of sleep is that it gives us a window into one of the great remaining questions in biology. Sleep is a state of altered consciousness. How does it differ from unconsciousness after a head injury or during anesthesia? And ultimately, what is consciousness, and how are the mind and the physical body related?

In this chapter, we will explore this key activity. We will review what sleep is, how sleep benefits us, which animals sleep and how sleep evolved, and some of the disorders that involve sleep. We'll start with a basic question.

What Is Sleep?

We all enjoy a good night's sleep, and we suffer without one. Lack of sleep is a serious problem in the United States (NHLBI, 2022). Nearly 40% of adults fall asleep during the day at least once a month, and 50–70 million Americans have sleep problems. Sleep deficiency leads to injuries and mental health problems and even increases the risk of death. It interferes with work, school, driving, and social interactions. Worse still, it is also associated with heart disease, kidney disease, high blood pressure, diabetes, stroke, obesity, and depression.

That seems like an easy question. But it isn't. Sleep is usually defined in behavioral terms. It generally features a rapidly reversible state of immobility and greatly reduced sensory responsiveness. Others have suggested behavioral and electrophysiological characteristics (Keene and Duboue, 2018). In a behavioral sense, (1) sleep involves a period of quiescence. (2) It is reversible by stimulation. (3) Each species has a specific posture while sleeping. (4) It takes more to arouse us when sleeping. (5) We experience a sleep rebound. Various electrophysiological measurements have been introduced to supplement the behavior characterizations. Importantly, sleep is carefully regulated so that when we lose sleep, we need to make it up, and after doing that, we usually experience a "sleep rebound."

Environmental factors can influence the quantity and quality of our sleep. Humans are essentially diurnal animals. We are active in the day and sleep at night. However, the modern world has made it possible for us to light the night, and many people must work or be active in other ways at night. Travel across time zones (jet lag), changes from daylight-savings to standard time and back, and other factors affect our sleep cycles. We try to manage the disruptions with alcohol, melatonin, chamomile tea, and other drugs. Caffeine, delivered in coffee, tea, or other drinks, is used by most Americans: 75% of US adults drink coffee, and 49% drink it daily (Loftfield et al., 2016).

Our need for sleep is related to the circadian clocks within our body (see Chapter 14) that generally recognize daylight and dark. Those clocks tell us when we need to sleep. They respond to sunlight, but they can also be influenced by artificial factors (electric lights) or caffeine. A recent study in the United Kingdom examined sleep in more than 100,000 people (Katori et al., 2022). It used the data from smart wristbands that monitored wakefulness, sleep, and arm movements during sleep. With the results, the authors could divide sleep into 16 distinct categories that easily sorted into 5 groups that may prove helpful in devising therapies for sleep disorders.

In more simple and classic terms, our brains have three states of being: wakefulness, REM (rapid eye movement), and non-REM sleep (Carskadon and Dement, 2011). The brain is active in all three states, but in different ways. Wakefulness is what we experience in our normal everyday life. Sleep and wakefulness are controlled by neurotransmitters and hormones. The neurotransmitters gamma-aminobutyric acid (GABA) and adenosine and the hormone melatonin help us to sleep. The actions of these biochemicals are antagonized by other factors. Light limits the production of melatonin and enhances levels of cortisol. Caffeine blocks the receptors for adenosine. These actions encourage wakefulness. The stress hormones norepinephrine, adrenaline, histamine, and cortisol also discourage sleep.

As we sleep, we cycle through REM and non-REM sleep several times each night and might even wake up in between them. Non-REM sleep involves three stages. We begin to fall asleep, and we enter a state of light sleep so that our breathing, heart rate, muscle activity, and brain activity slow down, and body temperature drops. Finally, we transition into a deep sleep state. Our heart rate and breathing are at their slowest point. It's difficult to wake us up at this stage. Scientists believe that the day's events are processed at this stage. REM sleep is different. As its name indicates, our eyes move around quickly, even while closed. Brain activity, breathing, and heart rate all accelerate. Dreaming occurs, but our nerves are paralyzed so we do not respond physically. Memories continue to be processed and stored. This phase of sleep occurs typically later in the night and early morning. Hayat et al. (2022) collected data from the deep brains of patients undergoing other procedures. They found that sound is detected by the sleeping brain in much the same way as it is in awake individuals, except that the brain cannot focus on it and so is not really aware of it.

How Much Sleep Do We Need?

We all like to get a good night's sleep, but the need for sleep differs for every individual. The average for adults is 7.5–8 hours per night, but individuals vary from 4 to 10 hours. Early on, we need a lot more. Newborns need to sleep 14–17 hours per day, and that number decreases as children grow up.

Interestingly, we sleep less than our primate relatives. Chimpanzees sleep about 9 1/2 hours each day. Cotton-top tamarins sleep 13. Three-striped night monkeys sleep an amazing 17 hours. In a study of 30 primate types, Nunn and Samson (2018) found that humans sleep a smaller fraction of a 24-hour day than other primates. Furthermore, they found that we have a higher proportion of REM sleep and that we have that because we have less non-REM sleep. They speculate that humans reduced their sleep when we moved from an arboreal to a more terrestrial existence. Sleeping on the ground increases vulnerability to predators. Also sleeping less gave us more time for learning and socializing. Humans in nonindustrial societies are similar to those of us in the industrialized world. We all sleep about 7 hours each day.

Many people are sleep deprived. For example, if the alarm clock wakes us before we would wake up ourselves, then we are sleep deprived. If we tend to fall asleep in periods that lack stimulation (boring lectures and books), that is another indicator. People lose sleep for several reasons. Too much noise can make it difficult to sleep. Watching television or staring at a computer screen bathes us in light that encourages wakefulness. Drinking caffeinated beverages also results in difficulty in sleeping. It is also difficult to sleep outside our normal circadian rhythm (during the day). Sleep disorders (e.g., apnea, insomnia) interrupt sleep. Some have suggested that it is important for allowing the body to recover and rebuild itself. That is particularly true of the brain. Sleep might help us to reorganize after the day. One function assigned to sleep is to consolidate newly acquired information into the memory (Diekelmann and Born, 2010). In times of sickness or injury, we might need additional sleep. Some have suggested that sleep allows us to divert energy from our normal activities to combatting an infection or repairing injured tissues (Prather et al., 2015).

Sleep Disorders

A good night's sleep is beneficial in many ways, but like so many things in biology, sleep sometimes goes wrong. Many people have trouble sleeping. *Forbes* magazine notes that Americans spend nearly $65 billion each year on sleep aids, and that number rises to $432 billion per year when beds, pillows, other sleep-related items are added in (Roberts, 2022). Yet, in the United States in 2020, 8.4% of adults used sleep medications every day or most days (Reuben et al., 2023). More women than men used the medications. Sleep is the second only to pain as the most common reason for people to consult a health provider (Mahowald and Schenck, 2005). The most common sleep complaint is insomnia. About 100 sleep disorders have been defined. They involve trouble falling or staying asleep, excessive sleepiness even when awake, circadian rhythm disorders, and behaviors during sleep. As with many fields of biomedicine, much can be learned by examining disorders.

Insomnia

Insomnia is the most prevalent sleep disorder that occurs by itself or with other medical disorders (e.g., pain) or psychiatric disorders (e.g., depression). The numbers are huge. About a quarter of adults complain about their sleep, 10%–15% have insomnia that affects them in the daytime, and 6%–10% have clinical insomnia (Morin and Benca, 2012). The symptoms include trouble falling asleep, staying asleep through the night, or waking up too early in the morning. The causes include lifestyle habits (e.g., an irregular schedule, napping, noise or light in the bedroom, lack of exercise, television or computer in bed), medicines or drugs (e.g., alcohol, smoking, caffeine, sleep and cold medicines), depression, stress, pregnancy, physical pain, waking to use the bathroom, and sleep apnea.

Snoring

Snoring is a common problem that affects nearly everyone at some point (Counter and Wilson, 2004). It is a harsh sound caused by the vibration of the relaxed tissues of the throat as inhaled air moves past them. In most cases, it is a nuisance for the victim or their partner. It can often be controlled by lifestyle changes, such as losing weight, avoiding alcohol before sleep, and sleeping on one's side. In other cases, it can contribute to more serious conditions.

Sleep Apnea

Sleep apnea can be serious, even life-threatening (Dempsey et al., 2010). It occurs when breathing stops and starts repeatedly. The symptoms include heavy snoring and being tired even after a night of sleep. Obstructive sleep apnea is caused by the relaxed tissues in the throat blocking the passage of air. Central sleep apnea is caused by inadequate control of breathing by the brain.

Sleep Terrors

Sleep terrors affect up to 6.5% of children and a smaller number of adults (Leung et al., 2020). They are not nightmares. The person wakes up from a nightmare and might remember some of it. A person with sleep terrors is still asleep. They may scream, thrash about, or run out of the room. They cannot be consoled during the event and typically do not remember anything after it. Fortunately, they are relatively harmless, and the person usually outgrows them. They only need attention if they are affecting the victim's sleep excessively.

Narcolepsy

Narcolepsy affects about 1 in every 2,000 people (Scammell, 2015). People with narcolepsy find it difficult to stay awake during the day. In fact, they fall asleep without warning, and so, this can be a very serious condition. Most patients also suffer from cataplexy, the sudden loss of muscle tone. They may also have sleep paralysis and hallucinations. Unfortunately, there is no cure for narcolepsy, although it can be partially controlled by medications. The cause is unknown, but patients have low levels of a chemical in the brain called hypocretin, which controls the onset of REM sleep normally. Several hypotheses have been proposed to explain narcolepsy, including genetics, loss of hypocretin to an autoimmune condition, or infection by a particular strain of influenza virus.

Restless Legs Syndrome (Willis-Ekbom Disease)

Restless legs syndrome causes unpleasant sensations in the legs so that the individual has a strong desire to move their legs (Allen, 2015). Up to 10% of the US population may suffer from it. Three clinical features define restless legs syndrome: a specific phenotype, low levels of iron in the brain, and involvement of dopamine. It has a strong genetic component. There is no cure, but it can be treated with iron supplements, antiseizure drugs, dopaminergic agents, opioids, and benzodiazepine.

Sleepwalking (Somnambulism)

Sleepwalking is often associated with another disorder (e.g., upper airway resistance syndrome, restless legs syndrome, sleep apnea) (Guilleminault et al., 2006). Treating those disorders can help with sleepwalking. It's more common in children, who usually outgrow it. The individual may sit up in bed with eyes wide open, but they do not respond to others, and they can be difficult to rouse. They can also walk around, leave their house, and even try to drive a car or engage in other activities. The risk of sleepwalking is that the patient can injure themselves. The cause is not really clear, but sleepwalking occurs in the deepest phase of non-REM sleep. Genetics and young age are two risk factors.

Do All Animals Sleep?

The answer seems to be yes. All of the animal species examined have some sort of sleep or sleep-like behavior. A more complicated answer depends on how one defines sleep (Siegel, 2008). We usually think of the sleep that we experience and involving periods of REM and non-REM sleep, little activity, and reduced responsiveness. The characteristics of human sleep were discussed above.

Everyone with a pet dog or cat has watched them running and moving while they are sleeping. Cats are notorious for sleeping long hours (12–16 hours) every day, but dogs are close behind (12–14 hours). They are typically most active just before dawn and at dusk. Some observers have speculated that that pattern correlates with wild cats looking for birds in the morning and rodents in the evening. Many of us who have had cats as pets have watched them moving and vocalizing in their sleep. They seem to be dreaming about hunting or other activities. Are they really dreaming or are we just projecting our human characteristics onto them? Like humans, cats cycle through different sleep stages, including REM and non-REM sleep (Peever and Fuller, 2017). They begin with non-REM sleep. During this light sleep state, they are ready to regain alertness immediately. They might go through a few additional phases of being alert, drowsy, and non-REM sleep before drifting into REM sleep. In this phase, their eyes are closed but often moving, their body might be limp or their muscles might twitch. Márquez-Ruiz and Escudero (2008) carefully characterized the movements of the eye muscles in cats during REM and non-REM sleep. They found similarities to those movements in humans.

However, not all animals display all those characteristics, and some animals have what seem to be variations on sleep (e.g., estivation, hibernation). Are those periods of relative inactivity similar to sleep? Methodologies used to measure sleep in humans are significantly limited in animals. Other aspects are even impossible to measure, especially in animals much lower on the evolutionary scale (e.g., fish, invertebrates). In many cases, scientists are left with behavioral observations. Observations that are consistent with sleep increase as we move up the evolutionary ladder. Some activities in single-cell organisms show a circadian rhythm. So do insects. Fruit flies have behaviors that seem to be sleep, but no one has reported REM sleep in insects. Fish also seem to have periods of rest that some refer to as "sleep swim." Fish, amphibians, and reptiles display some behaviors that seem like sleep. The evidence in reptiles is controversial. Birds have been reported to show both REM and non-REM sleep and to have behaviors associated with sleep. Land animals, mostly domesticated animals, seem to sleep. The amount of sleep varies considerably. Elephants and giraffes sleep little. Other animals sleep a lot.

One might also wonder how hibernation, brumation, and estivation fit into the concept of sleep by animals. Each of these terms refers to a strategy by an animal to survive harsh conditions by slowing their activity and metabolism

(Wilsterman et al., 2021). Hibernation is a form of deep sleep that is used by some warm-blooded animals in very cold conditions. It involves a low body temperature, slow breathing, slow heart rate, and a low metabolic rate. Bears, rats, and some birds hibernate during the winter. Some animals use a similar strategy for very short periods (<24 hours). This is called daily torpor. Brumation also occurs in the winter, but it involves cold-blooded animals, such as lizards and amphibians. Brumating animals may occasionally wake up to look for water or food. Deep hibernating mammals do not. Estivation is a strategy by animals to survive harsh hot and dry conditions. The animals lower their breathing, heart rate, and metabolism. Many species use some form of estivation.

The animals most familiar to us, typically mammals or at least vertebrates, show signs of sleep. Campbell and Tobler (1984) completed an extensive literature survey of 150 animal species, including invertebrates, fish, amphibians, reptiles, birds, and mammals. They noted that it is difficult to determine sleep in all cases. Siegel (2008) also examined sleep in multiple species.

Mammals and birds have well-documented REM and non-REM periods of sleep, although the periods tend to be shorter in birds. In addition, their behaviors are well documented and easy to observe. Behavioral observations show that reptiles clearly spend significant periods (6–12 hours per 24 hours) asleep. Electroencephalographic evidence supports these observations. Some researchers have also reported REM-like activity in reptiles, but Siegel (2008) found none in turtles. Reports on amphibians are mixed. The bullfrog *Rana catesbeiana* displayed periods of activity and inactivity, but interestingly, they were more responsive to stimuli in the inactive periods. Tree frogs (*Hyla septentrionalis*) were found to clearly sleep. Neither species show signs of REM sleep. Only a small number of fish species have been examined out of the approximately 30,000 species of fish, and so it is not clear yet that fish sleep. Obviously, it would be difficult to attach sensors to detect electric activity in the fish brains so sleep in fish has had to be tested behaviorally. The data on fish are sketchy, but there are some behaviors that suggest that at least some species undergo a quiet period each day. Others seem to swim continuously. Could they be sleeping or resting while sleeping? That is not clear, but it does occur in multiple marine mammals. In addition to sleeping on the surface and a "quiet hanging behavior," they swim while resting as if they go on autopilot for a while. Perhaps some fish also use this as a form of sleep.

However, all of those are vertebrates, and they are relatively easy to study. Invertebrates are the most difficult group to deal with for technical reasons. Many invertebrates have far less developed nervous systems than higher organisms, and most are aquatic. Nevertheless, several studies have provided interesting insights. For example, studies of circadian rhythms in insects have detected regular differences in activity in multiple species. A few other invertebrates also show periods of quiet and activity. Despite the difficulties, continued studies of these fascinating organisms are critical to our understanding of the evolution of sleep. Thus, the standard measurements of sleep (e.g., EEGs) are not possible with invertebrates and other organisms. Extrapolations from more imaginative tests for sleep are needed for other animals. For example, many studies have relied on behavioral tests. In addition, the powerful genetics of these organisms have enabled searches for genes involved in circadian rhythms and the sleep-like behaviors of the models. Several of those genes are analogous to the same genes in mammals. Some classic genetic small animal models have been examined, including fruit flies (*Drosophila melanogaster*), nematodes (*Caenorhabditis elegans*), and zebrafish (*Danio rerio*) (Keene and Duboue, 2018). Freiberg (2020) suggests that sleep evolved as a means for all life to adapt to the light and dark phases that occur every day. Evolution requires some sort of pressure. Freiberg believes that it is not possible for any organism to be able to adapt each day to two different environments. He concedes that the other arguments for sleep (e.g., repair, restoration), but he poses sleep as a mechanism to allow organisms to adapt to two rapidly changing environments.

The key behavior for the lower species is the same as it is for humans (for an excellent review, see Anafi et al., 2019). During sleep, humans exhibit reduced responses to stimulation, and after their sleep has been disturbed, humans need more sleep. Cockroaches (*Leucophaea maderae*), honeybees (*Apis mellifera*), and fruit flies (*Drosophila melanogaster*) show both characteristics. Even nematodes (*Caenorhabditis elegans*) show some evidence of a sleep-like state during development. For a few hours, they stop feeding and move and respond less. Thus, even an animal with only 302 total neurons may sleep. More interestingly, jellyfish (*Cassiopea* spp.) have more pulsations during the day than the night, and they lack a central nervous system altogether. In addition, some biochemical evidence also suggests sleep in lower animals. For example, melatonin promotes sleep in flatworms and caffeine promotes wakefulness in fruit flies.

Anafi et al. (2019) concluded that some type of sleep behavior existed at the time the Cnidarians (e.g., *Cassiopea*) split from the bilaterians (e.g., flatworms, mollusks, nematodes, humans). That event occurred about 600 million years ago. Thus, sleep seems to have an important function for the great majority of animals. It is not really surprising that sleep would have evolved early on. Sleep or at least sleep-like behaviors are found in many animals. Yet some remain hard to fathom. Sponges (phylum Porifera) are very simple animals. They have some diurnal rhythms, but it is still unclear whether they sleep. To extend the question further, do microorganisms sleep? The microbiome has become the object of much research into how it affects the whole organism (see Chapter 5). The composition of the mouse microbiome changes with sleep deprivation (Poroyko et al., 2016). However, it isn't yet clear whether these changes are related to anything that could be called sleep.

J. Allen Hobson (2005) stated that "Sleep is of the brain, by the brain, and for the brain." His view is supported by a great deal of evidence. Much of the study of sleep centers on brain waves during sleep and REM and non-REM sleep. Sleep deprivation causes a severe degradation of brain function. Sleep in vertebrates and other organisms is directed to a great extent by the brain (Saper et al., 2005). While there is considerable evidence for this point of view, not everyone agrees. For example, Vyazovskiy and Harris (2013) argue that individual neurons need sleep as much as the whole organism. They suggest that non-REM sleep provides time for those overworked individual neurons to recover in the same way that individual muscles need to recover and repair after hard workouts. Neurons are certainly the star players in sleep, but other cells might also have supporting roles. Other genes in other cells have been implicated. In mice, the brain and muscle ARNT-like factor 1 (BMAL1) in muscles influences sleep, and in fruit flies, NF-kB in the fat body (a liver-like organ) and DAF-16 in the muscles also have a part.

Do Plants Sleep?

On the one hand, it seems far-fetched. However, we often refer to flowers "going to sleep for the night" when they close their flower petals. Also plants do have circadian rhythms (Creux and Harmer, 2019). The primary ones are the dependence on the sun to provide energy through photosynthesis and the revision to respiration at night. Furthermore, sleep and the need for some form of sleep are shared by all animals that have been studied. Even many invertebrates seem to have a need for sleep. So one might wonder whether plants also have a need for some type of rest or recuperation.

In the very broadest sense, there is some circumstantial evidence. Flowers closing their petals at night is a real phenomenon. The scientific name for the process is nyctinasty. When the sun goes down, the air cools. The bottom-most petals of some flowers grow faster than the upper petals and force the flower to close. Interestingly, plants that are pollinated by bats or moths open their petals at night. Stomata on some plants also close at night to limit loss of water. Tree leaves and branches move slightly at night. Zlinszky et al. (2017) used terrestrial laser scanning to carefully document small movements of trees at night. They concluded that the movements were not in response to light. Animals hibernate or estivate when conditions are harsh. Plants produce seeds that go dormant in response to high concentrations of the hormone abscisic acid (Sussman and Phillips, 2009).

In the end, the case is not convincing. Referring to plants as sleeping is probably a case of anthropomorphism. We see activities that are remind us of our own activities, and the easiest way to describe them to others is to assign a human characteristic to them.

How the Brain Works

Neuroscience is one of the most important questions left to modern biology. While there are plenty of unknowns in how the brain works, a lot is known. At the cellular and molecular levels, the nervous system includes specialized cells, biochemicals, changes in ion concentrations, electrical transmission, and the diffusion of neurotransmitters across the synapse. All of these are used to send messages from distal receptors to the brain and back again.

All of our thoughts, memories, and actions are controlled and transmitted by neurons, which are the main workhorse of the system. Neurons have a long axon and many dendrites that make connections to other neurons and muscles. Each neuron has connections with many, perhaps thousands of other neurons.

Neurons use electrical impulses to transmit signals along their axons. Ion concentrations differ across the neuron membrane. At a resting state, there are more sodium ions inside, and more potassium ions outside. The gradients are maintained by proteins that pump the ions out of or into the neuron. When a neuron is activated, ion channels along the axon open to allow. Sodium ions flow out, and potassium ions flow in. This depolarizes the membrane at that site. It also causes the sections of membrane adjacent to the depolarized section to depolarize so that the action potential is transmitted along the neuron until it reaches the end.

Neurons are connected to other neurons and muscle cells at a synapse. The synapse is actually a tiny gap between them. The electrical impulse reaches the presynaptic termination of the neuron. There neurotransmitters held in tiny vesicles are released to diffuse rapidly across the synapse, and they are picked up by receptors on the postsynaptic side of the synapse. Once bound, they initiate another electrical impulse that continues the signal. The biochemicals involved in transmission include dopamine, GABA, acetylcholine, glutamate, and others. In addition, drugs are used to treat disorders of the nervous system.

In addition, much is known about the various regions of the brain and what each controls. Some of that was learned by studying individuals with injuries to specific areas of the brain. An injury to one area might result in the inability to speak or to learn new things. These studies have been greatly expanded by modern imaging technologies that can peer into the brain. These include MRIs, CT scans, and more. The resolution of these methods has increased dramatically in recent years. Finally, animal studies have provided valuable insights into how the human brain works.

What we don't know is how all of this is coordinated to form memories and thoughts. All of those connections between neurons hold the memories, but we simply do not understand the "code."

Christof Koch (2018) described his work with Francis Crick on the neuronal correlates of consciousness. By this, he meant the minimal portion of the brain that is responsible for consciousness and thoughts. They eliminated various parts of the nervous system (spinal cord, cerebellum, much of the frontal cortex) that could be lost without effect on a person's ability to think. They concluded that the critical area is that part of the posterior cortex containing the parietal, occipital, and temporal lobes. They labeled this region as the "hot zone." This work implicates certain regions of the brain in consciousness, but more work will be needed to further define where consciousness resides in the brain.

What Are Dreams?

German organic chemist August Kekulé dreamed of a snake biting its tail and the image of a benzene ring of six carbon atoms linked alternately by three single and three double bonds. This story is one of three cited by Paolo Mazzarello (2000) in a paper in *Nature* subtitled "The Scientific Benefits of Eating Cheese before Bedtime." He also cited instances in which dreams helped scientists solve major problems. Nobel laureate Otto Loewi planned an experiment that showed that the heart is controlled by chemical transmissions, and Dmitry Mendeleyev dreamed how the periodic table should be arranged. Mazzarello cites only a small data set, but it supports one of the theories that dreams allow the brain to process the day's information.

We can remember our dreams and our nightmares sometimes. They are formed from images, thoughts, and feelings and occur in a more-or-less random manner. Dreams are simply not understood, but they have provided fodder for arguments and study for centuries. Several theories have been advanced. For example, dreams may allow us to process thoughts and emotions about the events of the day. They may provide the brain an opportunity to reset itself in some way. They may promote learning or have evolutionary value as practice in avoiding threats. Or they may have no meaning at all. They are often forgotten as soon as we wake up. In some cases, fragments of a dream can be recalled, and one can learn to remember more with some practice.

The purpose of dreams remains as hazy as our memory of the dreams themselves (Winson, 1990). Freud and Jung developed sophisticated theories about the psychological value of dreams as windows into our psyche, and artists have tried to capture their meaning. However, neuroscience is limited to noninvasive methodologies by ethical considerations. Most dreams occur during REM sleep. More dreams are unpleasant than pleasant. Anxiety is a common feeling associated with dreams.

Whatever their function, dreams are likely to remain a mystery until more powerful imaging techniques are available to study them.

Consciousness

Sleep is a completely normal event for most of us. We go to sleep every pretty much every day for our entire lives. We only notice it when we miss it or when we are looking forward to it after a hard day. We might remember a fragment of a dream or nightmare that reminds us that we dreamed the previous night. But for the most part, we simply sleep and move on.

But what is the difference between wakefulness and sleep? Is dreaming a form of awakeness when we are asleep? And what about daydreaming? Windt (2021) proposed a continuum of conscious states. The continuum includes multiple levels of consciousness whether we are awake, dreaming, daydreaming, or not dreaming at all. Spontaneous thoughts and mind-wandering occur in both states. How can we deal with this range of consciousness? Windt does not completely answer that question, but she does point out that even considering it causes the gap between awake and asleep to be closed.

The continuity hypothesis (Horton, 2017) suggests that there is continuity between awakeness and asleep: our awake experiences and emotions carry on into our dreams in some manner. Yet sleeping is not homogeneous, as we have learned above, and measurements of sleeping experiences are problematic.

Many aspects of sleep are well documented. The health benefits of sleep are clear (Walker, 2008). The harm of sleep deficits is equally well known. This is perhaps the most fascinating aspect of sleep. It provides a window into the relationship of the body and mind.

However, the relationship between consciousness and the physical brain is one of the great remaining questions in biology. Understanding this is the ultimate goal of neurobiology, but there is much work to be done to tie together ion fluxes, electrical currents, axons and neurites, and thoughts and memories. Sleep is one facet of the problem, but we can also wonder about its relationship with consciousness, anesthesia, and coma.

While we are awake, the brain is receiving inputs from our senses: sights, sounds, smells, tastes, and sensations of touch. Our brains take that information and construct a model of the world. But as we fall asleep, the input from our senses diminishes. When we sleep, we exist only in our own mind. The brain is left alone. Interestingly, it is busy. In fact, REM sleep has as much activity as waking. Although we have some dreams during non-REM sleep, most dreams occur during REM sleep. We spend about 2 hours each night dreaming, and each dream lasts 5–20 minutes, although they might seem longer or shorter to us after we wake up. We usually only remember snippets of a dream. In REM sleep, few external stimuli are being processed, and our muscles are at their most relaxed. Yet, imaging experiments have shown that brain areas associated with visual and motor experience and emotions are as active in REM sleep as in waking (Desseilles et al., 2011).

We have few tools to probe the workings of dreams. Ethical considerations prohibit the use of invasive technologies in most cases. The exceptions are those rare cases where invasive techniques can be used in parallel to a procedure needed by the patient. Thus, we are limited to observing from the outside. For example, electroencephalographs and other imaging methods allow some insight into what is happening in the brain. Nevertheless, these methods are rapidly increasing in power and resolution and are likely to provide better models in the future.

Loss of Consciousness

An unconscious person lacks any knowledge of themselves or others and is unresponsive to stimuli. More technically, a loss of consciousness involves a loss of the ability of the brain to integrate bits of data from our senses and memories into a coherent whole or information. Sleep is one form that we are all familiar with, but there are others that are far less benign; these might result from head trauma, brain hypoxia, intoxication, pain, or anesthesia. Coma is a state of profound unconsciousness. A person in a coma does not respond even to painful stimuli and may have limited control of various bodily functions. In some cases, a person can revive from a coma and regain most or all of their cognitive abilities. In other cases, the person can be in a vegetative state. Those individuals have suffered severe trauma to the cerebral cortex and will never come out of the coma.

Anesthesia

For nearly 200 years, various anesthetics have been used to induce a sleep-like state during surgeries. The many different anesthetics have different mechanisms and targets in the brain and spinal cord. In general terms, they control synaptic transmission and membrane potentials by interacting with ion channels. In this way, they block pain messages to the brain. At low doses, anesthetics induce analgesia, amnesia, and sleep. At higher doses, the patient does not respond, and they are considered ready for the procedure. Yet, are they actually unconscious?

There are two broad types of anesthetics (Alkire et al., 2008). Intravenous agents are used for anesthesia, typically with sedatives or narcotics. Volatile agents are used to maintain anesthesia. They act by regulating synaptic transmission in the brain and spinal cord. Most anesthetics seem to target a specific region of the brain called the posterior lateral corticothalamic complex.

Verbal commands are often used to determine whether a patient is unconscious. However, this might not be adequate. The patient can seem to be asleep, but it isn't quite the same. The patient might be unable to move or speak but still conscious. The anesthetics have different modes of action and affect different parts of the brain. Those also differ with the dosage.

Sleep and anesthesia have some common elements, but also some differences. In both, brain arousal systems are inactivated. Both result in slow wave cycles of brain activity and an inability to integrate information.

Conclusion

The influence of the light and dark of a typical day likely formed patterns very early in evolution. For example, an active period in the light and a quiescent period in the dark would have offered the opportunity to use that downtime for rest and repair. In that sense, it answers the question posed at the beginning of this chapter by Dr. Walker. It must have "an absolutely vital function," and so, it is worth the risk in evolutionary terms.

14

Telling Time

The author proposed to alter the time of the clock at the equinoxes
so as to bring the working-hours of the day within the period of daylight and, by utilizing the early morning, so reduce the excessive use
of artificial light which at present prevails.

—George Vernon Hudson (1895)

Biologist George V. Hudson (1895) was likely disappointed by the reception of his recommendation. It was not immediately acted on. He was
hoping for extra daylight hours to allow him to collect more insects in New
Zealand. Still, various countries played with the idea of daylight savings time
throughout the 20th century. The United States used it during the two world
wars and then officially adopted it in 1966. Now twice each year, we in the
United States change to daylight savings time from standard time and back
again, and we feel the effects for several days thereafter. In theory, the changes
are a good idea to maximize the amount of daylight that we have available for
activities. In practice, it has some costs.

Time was not always so carefully measured for humans. Early hominids
and later humans set their days by the sun. Only relatively recently have we
become more keenly aware of time. The Babylonians and Egyptians began
using calendars to track of community events (Andrewes, 2006). In about
2000 BC, the Sumerians devised a system of time based on a 60-minute hour
and a 60-second minute. Reasonably reliable clocks have only been around
since the end of the 13th century. The first clocks were mechanical. In 1283,
a weight-driven clock was invented at Dunstable Priory in Bedfordshire,
England. In 1656, Christian Huygens invented the pendulum clock. Before
that, there were sundials, water clocks, candle clocks, and hourglasses. Our
ability to measure time now is extraordinary.

Today we are controlled by time. Our alarm wakes us each morning to we
get ready to get to work or class at a specific time. The day is filled with times

The Biology of Us. Gary C. Howard, Oxford University Press. © Oxford University Press 2024.
DOI: 10.1093/oso/9780197664797.003.0014

for classes, meetings, lunch, coffee breaks, phone calls, and more. Then we go home to dinner and to watch a game or other show on television. Airplanes and commuter trains are scheduled to the minute. We have appointments. And we always know the time from our ubiquitous cell phones. Yet, we sometimes are working against ourselves.

Time is also important in nature. The sun rises and sets on an approximate 24-hour cycle. The moon waxes and wanes over about 29½ days. The Earth orbits the sun in about 365 days. Throughout evolution, these rhythms have become ingrained into most living things. Plants and animals sense the changing seasons as signals for when to reproduce, migrate, or hibernate. The phases of the moon that control the tides also govern the behavior of multiple marine species. We humans respond to the rising and setting of the sun.

Circadian Rhythms

Our internal clocks regulate many aspects of our physiology and behavior. The Earth rotates approximately every 24 hours, and those 24 hours constitute a day, and thus, our circadian rhythms are based on a 24-hour cycle. We humans, as well as most organisms, evolved our physical, mental, and behavioral activities to fit the light and dark periods of that clock. For the most part, we humans sleep at night and are active during the day.

Interestingly, humans are born without a circadian clock. In a way, it is not surprising. The womb is dark. Newborns "learn" the 24-hour cycle within about the first 4 months of life. Body temperature is important, and so, infants develop a cycle that involves spikes in temperature right around falling asleep. Cycles for melatonin and cortisol begin in those months, and those hormones assist the newborn in establishing the needed cycles.

The master clock begins with light. Input from the eyes is important to our biological clocks. Light is detected by our eyes. Special neurons at the back of the eye called intrinsically photosensitive retinal ganglion cells (ipRGCs) detect the light (Pickard and Sollars, 2012). They are separate from the normal rod and cone retinal cells that we use for our vision and contain a different photoreceptor, melanopsin. The cells send signals via the optic nerve to the suprachiasmatic nucleus (SCN). This complex contains about 20,000 neurons and lies in the hypothalamus region of the brain. Thus, the eyes supply information directly to the master clock.

The rhythmic nature of various biological processes has been known for some time. We and most living organisms are kept on schedule by biological clocks that involve specific genes and proteins that interact with specific cells to direct those cells to do things. The genes that encode those proteins are conserved across many species, including humans, flies, mice, fungi, and plants. The system comprises a complex hierarchy of peripheral mechanisms at the cellular, tissue, and systems levels (Mohawk et al., 2012). The SCN provides the central control for the subordinate clocks. However, the peripheral clocks maintain a circadian rhythm even without input from the SCN. Indeed, each cell seems to have its own internal clock that is under genetic control so that they continue on their own with a kind of "quality assurance" from the SCN. Various genes and gene pathways control the function of internal clocks. This system of clocks is intimately involved in multiple physiologic functions. Thus, it is not surprising that the disruptions affect health and disease. Interestingly, those disruptions can also affect the how well some therapies work and whether they are toxic under certain circumstances (Ayyar and Sukumaran, 2021).

Each of these is involved in controlling various physiological processes that are related to the light/dark cycle of a normal day, and the most obvious of those is sleeping and waking. However, circadian rhythms control much more than just sleep (Reddy et al., 2023). They control body temperature, and levels of blood glucose, catecholamines, insulin, and hormones (e.g., melatonin, cortisol, TSH, ghrelin, leptin and prolactin). They also regulate the genome, epigenome, metabolome, proteome, and microbiome (Skarke et al., 2017). Circadian rhythms control translation and transcription and regulate various signaling pathways and posttranslational modifications, such as phosphorylation.

In 1971, Konopka and Benzer (1971) discovered mutants in fruit flies (*Drosophila melanogaster*) that causes the flies to have unusual circadian rhythms. Instead of the normal 24-hour cycle, these mutants had cycles of 19 and 28 hours, and one mutant had no rhythm at all. Konopka and Benzer were studying eclosion, which refers to the process by which flies emerge from their pupae as adult flies. This normally happens at dawn, but the two scientists isolated mutants with shorter and longer periods or no period at all. They put flies on a 12-hour light/12-hour dark cycle and moved pupae into a dark environment at the end of a light period. They then took samples of the pupae every hour to look for those beginning eclosion. With this strategy, they found the mutants. The mutations were linked to the X chromosome, and the scientists named the gene *period* (*per*). Other scientists later isolated the product of the *per* gene, the Per protein. Later other genes were found with roles in maintaining circadian rhythm. One example is a mutation in a gene in flies called *timeless* (*tim*) (Vosshall et al., 1994). The control of our sleep cycle also involves a protein called cryptochrome 2 (cryptochrome 1 is only found in invertebrates). This blue-light receptor was discovered in plants, but is found in essentially every organism. In mice, a mutation in this gene shortens the circadian period and interferes with sleep (Hirano et al., 2016).

The cycle in the cells is complex, and it features a large cast of characters that are known mainly by their abbreviations. Here is a summary of the processes. The cycle is overseen by CLOCK and BMAL1 (Mohawk et al., 2012). They turn on Per and Cryptochrome (Cry) at the beginning of the cycle. As noted above, Per and Cry bind to each other (dimerize) and cross the nuclear membrane to turn off CLOCK and BMAL1. The levels of Per and Cry are controlled by an enzyme that attaches a ubiquitin to them. Ubiquitin is a small protein that signals that the protein it is attached to is ready to be recycled. Other feedback loops are involved, and the whole cycle takes about 24 hours. For example, the Per protein is made by the cells during the night. As it accumulates, it binds to the Tim protein. Once bound, the dimer can enter the nucleus and cause the *Per* gene to stop making Per mRNA in a negative feedback loop. During the day, Per protein is slowly degraded. By nighttime, the supply of Per is exhausted, and the whole process begins again. By the 1980s, new molecular genetic techniques enabled Jeffery Hall, Michael Rosbash, and Michael Young to isolate the gene. For this achievement, the three scientists shared the 2017 Nobel Prize (Sehgal, 2017).

Circadian clocks are not limited to animals. Plants also clearly react to the light/dark cycle of a normal day. The control of photosynthesis is critical in plants. As the sun sets, they stop photosynthesis and begin to rely on respiration. Some flowers close their petals to "go to sleep" at night and open them to "awaken" in the morning. Stomata are pores in the leaves and sometimes in other plant structures. They close in the evening and open in the morning to control gas exchange and retain water. The feedback loops control the opening and closing of the stomata, which control the movement of carbon dioxide into the plant. The rate of synthesis of chlorophyl is also controlled by the circadian rhythm genes. As a result, the production of NADPH and ATP varies throughout the day.

Plant circadian activities are also controlled by a central oscillator with three feedback loops (i.e., morning, central, and evening) (Venkat and Muneer, 2022). In the 1930s, Erwin Bunning noted that plant leaf movements were controlled internally by genetics (Pittendrigh, 1993). Each leaf has genes that encode various transcription factors that, in turn, regulate other genes in their loops. Much of the experimental work on circadian rhythms in plants has been done on thale or mouse-ear cress (*Arabidopsis thaliana*). In this plant, the cycle can be lengthened by overexpressing the transcription factors BBX18, BBX19, and BBX32 (Venkat and Muneer, 2022). The plant internal clock is involved in the regulation of metabolism, development, flowering, and senescence.

Confusing Our Internal Clocks

As noted above, circadian rhythms control our hormones, eating habits, body temperature, and sleep patterns. Several common situations can disrupt our internal clocks. When they are thrown off, we may have sleep disorders and other problems, such as obesity, diabetes, depression, bipolar disorder, and seasonal affective disorder.

Daylight Savings Time

As noted at the beginning of the chapter, our circadian rhythms are affected by the changes that occur when daylight savings time begins or ends. Many people feel groggy, tired, or not quite with it for a few days after. The change in the spring seems to be more severe. The anecdotal evidence is supported by multiple studies. Many people complain of disturbances to their sleep at that time (Johnson and Malow, 2022). Other complaints concern depression and foggy thinking. Those most affected tend to be adolescents, those who work irregular hours, and those who begin work early in the morning. Several studies show that traffic accidents increase for about the week after we leap forward in the spring (Prats-Uribe et al., 2018). Fritz et al. (2020) completed a very large study of fatal traffic accidents from 1996 to 2017. They found that fatalities increased for the week after the spring transition but returned to normal levels in the second week. The number increased even with more light in the evenings. There was no effect in during the "fall back" transition in the autumn. Manfredini et al. (2019) conducted a meta-analysis to determine the risk of acute myocardial infarction at the transitions. They found that a modest, but significant, increase in the risk of heart attacks. Thus, the time change seems to be related to a disturbance to circadian rhythms that is mostly inconvenient, but has some serious consequences.

Jet Lag

Jet lag results when a person's internal clock is out of sync with their environment. It usually happens after travel from one time zone to another. The effect seems to be greater when the day is shorter than that to which one is accustomed. In other words, it is more difficult to travel from west to east than vice versa. The major symptoms are fatigue and difficulty sleeping, and these usually resolve in a few days.

Some of the best data on the effects of jet lag come from Major League Baseball. Baseball has been keeping records for over 100 years. During the regular season, each team played 154 games until 1961, when the season was increased to 162 games. About one-half of those games were away games, and a significant number occurred after the team had traveled from coast to coast in either an eastward or westward direction. The performance of so many teams after so many coast-to-coast trips provides an amazing opportunity to examine the effects of jet lag. Games are played all around the country, and the travel time can be significant if a game is played in Boston and then the team has to fly to Los Angeles or, worse still, from LA to Boston (Winter et al., 2009; Song et al., 2017). Song et al. found a significant effect on homes runs allowed after jet lag.

In like manner, McHill and Chinoy (2020) compared records from the National Basketball Association before and after the "pause" due to COVID-19 for effects due to jet lag. They found that jet lag reduced the winning percentage, team shooting accuracy, turnover percentage, and more.

Screen Time

Light resets our sleep cycles. For most of our evolutionary history, that was no problem. There was no light at night. The coming of electricity and the incandescent bulb changed that. More recently, other challenges to sleep have appeared. Televisions became common in the 1950s, but today LED screens are ubiquitous. There are in our phones, computers, and televisions, and we spend enormous amounts of time staring into them. Moreover, they emit a lot of blue light, which is exactly the light that is most effective at activating ipRGCs. It can confuse those receptors so that they signal the SCN to turn off the production of melatonin. The controls on our circadian rhythms usually work very well. Unfortunately, they evolved a long time ago, and so, they are not able to serve us as well in the modern world.

Extreme Northern Latitudes

How do organisms at the extreme northern or southern latitudes deal with the extended hours of light in summer or darkness in winter? For example, reindeer (*Rangifer tarandus*) live at those extreme latitudes where there is essentially no dark in summer and no sun in winter. Are their internal clocks affected by the sun that never completely sets in the summer? Melatonin seems to not be involved. Lu et al. (2010) showed that the cell's internal oscillatory mechanisms must control sleep cycles in the reindeer. The authors speculate that there is another neural network that compensates for the loss of the traditional circadian rhythm mechanisms.

Melatonin

Melatonin is a naturally occurring hormone that helps us sleep (Gandhi et al., 2015). It is formed in the dark, and light inhibits its production. Many people take melatonin supplements to aid their sleep, and it might be helpful in some situations (NIH, 2022). The NIH (2022) cites several studies that suggest that melatonin helps with jet lag, especially on eastward flights. There is some evidence that it also helps children with some challenges, such as autism and attention-deficit hyperactivity disorder. Melatonin is also recommended for chronic insomnia. However, the effectiveness of melatonin in helping shift-workers sleep is much more controversial. Long-term studies on the safety of melatonin are lacking, but the good news is that short-term use seems to be safe. Use of any supplement, especially with children, should be discussed with a healthcare provider in advance.

Beyond Waking and Sleeping

Clearly, circadian rhythms are intimately involved with multiple facets of the lives of plants, animals, and other living organisms (Foster and Roenneberg, 2008). But there are other physical periodicities that we all experience. The Earth, sun, moon, tides, and more have influences on us (e.g., day/night, tidal, lunar, and annual). Tides cycle every ~12.4 hours. The moon rises and sets every ~24.8 hours, and a lunar month is 29.5 days. How do these physical changes help living organisms to tell time?

The moon has long been assumed to influence humans. For example, a full moon caused mania in people (Iosif and Ballon, 2005). Perhaps it is a remnant of times before lights at night. Whatever its cause, it has been a premise for many stories, novels, and movies. Ironically, studies on human behavior and physiology do not support the narrative. At best, they are controversial. At worst, there is simply no evidence at all. Still, it makes for great stories.

The moon may have no effect on humans, but it does influence the tides and many marine organisms. Horseshoe crabs (*Limulus polyphemus*) are influenced by water pressure, but light and temperature are less important. The phases of the moon affect many species. In particular, the timing of their reproductive behavior and physiology is controlled. For example, the intertidal midge (*Clunio marinus*) uses the moon to time its eclosion and mating. The moon might also affect hormones in insects and birds and reproduction in fish (Zimecki, 2006).

Many seasonal rhythms involve telling time. In the autumn, leaves and ripe fruit fall. As noted in Chapter 10, programmed cell death facilitates their separation from the branch, and gravity does the rest. Birds, butterflies, and other insects migrate. So do various herds. Some of them for very long distances. Perhaps they detect the shortening of the days, a reduction in the availability of food, or a chill in the air to prompt them to leave.

Development and Aging

Telling time is not always as obvious as the time of day, the movement of the sun, the phases of the moon and the tides, or the comings of the seasons. For some processes, the timing of events is important, and key examples of those are development and aging.

Development

Humans develop from a single fertilized cell to an organism of approximately 30 trillion cells arranged in many organs and tissues. Timing is absolutely critical. Mistakes result in birth defects. Development is controlled by many genes turning on and off at just the right time and in the correct sequence. Development occurs in stages. The genes code for RNAs (e.g., microRNAs) or proteins (e.g., transcription factors) that control further transcription or translation. Steroid hormones and programmed cell death are also involved (Moss, 2007). In some cases, certain tissues need to be removed as others are developing. Gradients of the transcription factors direct cell growth. To develop a human or other complicated organism is an extremely intricate operation, and timing is everything.

MicroRNAs are key elements in the regulation of development. These are small, single-stranded noncoding RNAs of 21–23 nucleotides. They regulate the expression of genes by silencing messenger RNAs and other ways after transcription. The nematode *Caenorhabditis elegans* has often been used as an experimental model because development in vertebrates is extremely complicated and difficult to unpack. Development in *C. elegans* is controlled by a cascade of regulatory genes and gene products. Two microRNAs, lin-4/mir-125 and let-7, are important for regulating gene expression in the development of the nematodes and other animals (Ambros, 2011). Let-7 is a heterochronic switch gene. That means that mutations in that gene cause developmental events to occur at an incorrect time. If let-7 is hindered, some cell fates are reiterated (Reinhart et al., 2000). If the level of let-7 is increased, the process is accelerated so that the adult phase occurs abnormally early. Let-7 seems to control several genes, and lin-41 seems to control other genes. Furthermore, these genes work together to ensure that development follows a predetermined pattern and the nematode develops appropriately. This is a small peak into a complex choreography that involves multiple genes and many proteins, hormones, and microRNAs.

Mammalian organ development is even more complex and largely unexplored. Cardoso-Moreira et al. (2019) studied the transcriptomes of seven organs in multiple species. Interestingly, they found similar timing of genes expressed across the species during organ development.

Aging

In some ways, aging is development in reverse. The timing varies, but it eventually begins. Gray hair, wrinkles, hearing loss, and aches and pains come on slowly but inexorably. We fight it with lots of money. We use concoctions to remove the wrinkles from skin, put color back into hair, and increase stamina and muscle tone. But it is in vain. Some physical changes threaten our vanity, and we laugh about them. But many are no laughing matter. They threaten our lives. Overall fitness declines. Bones lose mass (i.e., calcium and other minerals) and become more brittle (Boskey and Coleman, 2010). Digestion slows. The brain changes as we age (Harada et al., 2013). Tragically, these events are progressive and ultimately fatal, and no therapies are available.

Interestingly, the first organisms probably did not age. Since they were simple cells, they likely reproduced by equal cell division (Ackermann et al., 2007). The result was two equal daughter cells. The damaged proteins and other components would have been distributed roughly equally, and so, the two daughter cells would have been rejuvenated somewhat by the reduced number of those damaged components. Thus, those early cells did not "age."

That situation did not last forever. At some point, the daughters cells received different amounts of the damaged components. That cell with more would have a more difficult time continuing to live, but the one with less would have an advantage. This asymmetry might provide a mechanism for "aging" in bacteria and yeast (Watve et al., 2006; Clegg et al., 2014; Yang et al., 2015). If the damaged material could be segregated into one daughter cell, that mechanism would likely be favored by natural selection. In fact, some bacteria have been found that do exhibit asymmetric division.

There are plenty of theories to explain aging, and they are not mutually exclusive. The *mutation accumulation theory* states that genes that are deleterious later in life have less or even no negative selection. The organism has already reproduced so genes with later affects are essentially meaningless for evolution. A good example is Huntington's disease. This disease manifests late in life after mutations have accumulated. *Antagonistic pleiotrophy theory* posits that a single gene might affect multiple traits. According to the theory, aging results when short-term benefits, primarily those related to reproduction, are favored over long-term survival. Organisms are enabled to maintain themselves very well before reproduction, but afterward, mutations cause senescence and ultimately death. *Disposable soma theory* states that organisms have many requirements for their resources, and so, they sometimes must make compromises about how they allocate those limited resources. They must balance the investment of resources in reproduction or repair. If they spend too much on repair, their offspring will suffer and perhaps not reach reproductive age. Thus, reproduction is favored and receives most of the resources. With fewer resources available for repair, cellular damage is not fixed.

The three classic theories of aging provide a powerful basis for thinking about the evolution of aging. Recently, several newer theories or variations on the classic theories have been suggested. Like the classic theories, they are not mutually exclusive. They are also not completely compelling. *Reliability theory* points out that aging is a trade-off. Redundant systems (e.g., proteins) enable the organism to maintain good health early on, but later those systems fail. For example, the accumulation of damaged proteins has been implicated in several neurodegenerative diseases, such as Huntington's disease, Alzheimer's disease, and Parkinson's disease. *DNA damage theory* suggests that aging results from an accumulation of unrepaired damaged genes. *Wear and tear* says that machines simply wear out. Aging might be due to simple wear and team of an organism's systems as the repair mechanisms are overwhelmed. Lidsky and Andino (2020) propose that outbreaks of infections contribute to moderate the evolution of setpoints. Thus, in their theory, epidemics are intimately linked to life span. They note that organisms with shorter life spans experience a lesser spread of infection and clear pathogens faster than those with longer life spans.

None of the theories of aging is completely satisfying. There are always confounding aspects that can't be explained. For example, one confounder is the naked mole rat. They are a favorite model for aging. They live in colonies with only one breeding female and a small number of breeding males. Also they live a long time (37 years) and, most amazingly, do not age by the normal measures. They seem to be immune to most age-related diseases (e.g., cardiovascular disease, cancer). Their mortality rate does not increase with age as it does in most other organisms. Horvath et al. (2022) examined the naked mole rats for patterns of DNA methylation. This epigenetic markers, such as cytosine methylation, change with age. By surveying multiple methylations, an epigenetic clock can be established. Interestingly, the epigenetic clocks for naked mole rats are similar to those in other animals, but that of the queen ages more slowly.

Sex and aging are related. The exchange of genetic information during sexual reproduction is a great advantage in evolution. It provides additional diversity that allows the ability to change. Furthermore, maybe sex is the cause of aging. However, there are costs, and those costs differ between males and females (Brooks and Garratt, 2017). Both sexes compete for mates, but females bear the burden of gestation and lactation. Males through much of evolutionary history had only to provide sperm and defend territory. The different costs result in differences in aging.

Telomeres are another device that is involved in aging. The simplest image for a telomere is the plastic end that prevents shoelaces from fraying. In a similar way, telomeres protect the ends of chromosomes. These DNA-protein structures protect prevent chromosome degradation and fusion and, thus, are essential to maintain genomic integrity (Armanios and Blackburn, 2012). They also act as a molecular clock. Each time a cell divides, the telomeres become a little bit shorter. Over time and multiple divisions, they can become so short that the cell can no longer divide successfully.

Telomeres have been known for quite a while (Chakravarti et al., 2021). They were inferred to exist in maize by Barbara McClintock in 1931 and in *Drosophila* by Herman Mueller in 1938, and Mueller coined the term "telomere." The technical reason for the telomere's importance is that a linear DNA cannot be replicated completely. The DNA in chromosomes is linear and double-stranded. To replicate both strands, the process goes in opposite directions. The so-called leading strand is synthesized as a complete polymer, but the "lagging strand" is synthesized as a series of small Okazaki fragments, each of which has its own primer. This was known as the "end-replication problem," and it was hypothesized in the early 1970s by Alexei Olovnikov and James Watson. It posited that replication damaged the chromosome ends. Because of this shortcoming, the chromosomal DNA becomes shorter after each replication cycle. Telomeres also fitted well with a concept known as the Hayflick limit (Hayflick, 1965). That limit reflected the finding that human fetal cells could only divide for about 50–60 times before dying.

Importantly, telomeres provide a solution to this problem. They contain short repeated sequences that are mainly multiple tandem TTAGGG repeats. This DNA is structural and does not code for any proteins. DNA sequences that are rich in guanines tend to form secondary structures called G-quadruplexes (Antcliff et al., 2021). Four guanines use Hoogsteen hydrogen bonding to assemble a square planar structure called a guanine tetrad.

Each telomere contains a few hundred nucleotides of the repeats. At each replication, some of the telomere sequences are lost (Aubert and Lansdorp,

2008). The repair mechanisms involve the enzyme telomerase. Telomere length is maintained in the cells of the germline, but somatic cells typically lose some sequences at each replication. This mechanism prevents most cells from growing uncontrollably, but as a result, the telomeres of most cells shorten with age. When the number of chromosome ends lacking sufficient telomeres is reached, apoptosis or cellular senescence is activated, and the cell dies. Individuals with mutations in telomerase have a heightened risk for several diseases, including dyskeratosis congenita, aplastic anemia, pulmonary fibrosis, and cancer. The rate of loss of telomeres is increased by stress.

Time

Lewis Wolpert said, "I chose the most famous quotation in developmental biology: 'It is not birth, marriage or death, but gastrulation which is truly the most important time in your life.' What, we are invited to ask, is gastrulation? Answer: it is the processes early in animal development that turn a ball of cells into a three-dimensional body with the main parts in more or less the right places" (Hopwood, 2022).

Wolpert's quote sums up the importance of timeliness in biology. The field of chronobiology studies the relationship of time and biological processes. In fact, time is a key factor in many biological processes, and throughout evolution, organisms have developed the ability to match activities (e.g., metabolism, development, reproduction) to an optimal environment. Circadian rhythms are paramount because essentially all living organisms have matched their activities to the 24-hour cycle of the sun. However, other physical parameters govern various aspects of life. The phases of the moon, the ebb and flow of tides, and the changing seasons also signal the beginning of activities for various organisms. Internal clocks and other genetic and epigenetic mechanisms have evolved to enable organisms to optimize their existence. Mutations or other disruptions of the timing of these processes result in birth defects, cancers, and even death. Telling time, in all its manifestations, is critical.

15

What's Next?

The anthropologist Donald Johanson (Bohn, 2009) correctly noted that human evolution has not stopped. Our modern advances shield us from many evolutionary pressures. For most of us in the developed world, life is fairly easy. Our hunter-gatherer ancestors had to spend much of the day looking for food. Now we hunt for bargains at the local supermarket and gather whatever we want. We are protected from the weather. Our homes are heated in the winter and cooled in the summer. Yet, we are still part of biology and evolution. If we had any doubts about that, the COVID-19 pandemic should have ended them.

Life began in the sea, it is assumed, and at some point much later moved onto land (Schulte et al., 2015). Organisms had to develop mechanisms to deal with retaining water, balancing salts, dealing with gravity, gas exchange, nutrient distribution, and much more. Stephen Jay Gould (1990) wrote in his classic book *Wonderful Life*, "Life is a copiously branching bush, continually pruned by the grim reaper of extinction, not a ladder of predictable progress" (p. 35). Evolution is complicated with lots of twists and turns and many forks in the road. We humans are a leaf on the bush of life, and we are the products of billions of years of evolution that constructed the many systems that make us up over time. We depend on our collection of organs, tissues, bones, and other components to stay alive. Each of those has evolved to allow us to do that. Those mechanisms have been refined over millions of years of evolution.

In this book, we have looked at ourselves, how we humans fit into the rest of biology, and how we got here. We looked at the mechanisms for eating and digesting food, standing up and walking, the use of programmed cell death in development and defense against disease, communication, and timing of events. We looked carefully at some of those steps that differentiated us from other living organisms. Many of our bodily functions continue without any real awareness on our part. Most are so fundamental that they are controlled by the autonomic nervous system, so we do not even think about them.

The Biology of Us. Gary C. Howard, Oxford University Press. © Oxford University Press 2024.
DOI: 10.1093/oso/9780197664797.003.0015

Furthermore, we learned how to insulate ourselves from many of the insecurities of nature. However, even with all of our modern advances, we are still part of the living world. We are all still part of biology. It is all around us. We just need to look. We are also still part of the process of evolution. Multiple lines of evidence support this assertion. For example, Hawks et al. (2007) examined genetic markers from four groups (i.e., Han Chinese, Japanese, Yoruba, and northern Europeans). They found that 7% of human genes had undergone some degree of evolution within the last 5,000 years. The ability to digest lactose appeared in regions that drink a lot of milk. They also estimated that humans have evolved 100 times faster in the last 10,000 years than any time since humans split off from the great apes. Karlsson et al. (2014) found recent changes that provide resistance to diseases, including Lassa fever and malaria, and changes in skin and eye color. One measure of the changes to the human genome is the reduction in the percentage of Neanderthal DNA in modern humans that occurred over the last 35,000 years (Fu et al., 2016). Another particularly fascinating example involves eye color. Until about 6–10,000 years ago, all humans had brown eyes. However, a mutation occurred in a single individual, and everyone on Earth with blue eyes descended from that person (Eiberg et al., 2008). The mutation caused the gene for a protein that is involved in the production of melanin to be turned down but not off. The lower amount of melanin allowed the blue color to appear. A similar effect results in green eyes, but the amount of melanin is much lower in blue-eyed individuals, and the low degree of genetic variability in blue eyes indicates that it is a recent mutation.

Things that might have driven evolution in the past no longer do (Ward, 2009). International travel has slowed the isolation of different groups. Our modern life limits our exposure to predators. We are protected from much of the weather. Modern medicine has alleviated many diseases. Food is readily available. In fact, we are now awash in calories and experiencing the diseases of affluence. One might wonder if evolution has ceased. Or we might seek to direct our own evolution as we come to better understand human genetics. We have certainly done that with many other plants and animals. Such manipulations are considered unethical today, but the temptation is always there for someone willing to push the envelope. Some have already sought to select for the sex of children.

Artificial intelligence may also have serious implications for our future. We don't fully understand the possibilities and challenges of it as yet, but there are some hints out there. Some of those hints point to beneficial outcomes for those with spinal cord injuries and other injuries involving the loss of abilities due to nerve injuries (e.g., limb loss). But the possibility of mischief is very strong.

However, we are entering a new era of climate extremes, and we might learn the hard way that evolution is still active.

We share a common ancestry with every living organism on Earth. Amazingly, in a biological world in which untold numbers of species and individuals have lived and died, we right now are the survivors of 3.5 billion years of evolution, including at least five mass extinction episodes. Extinctions are not unusual. If anything, they are common, and they occur nearly continuously. For many different reasons, a species can no longer adapt to its surroundings, and it dies out. Perhaps 90% of all species that ever existed have been lost.

Mass extinctions are hard to determine. The difference between normal extinctions and mass extinctions is one of degree. They are defined as the loss of three-quarters of all species on Earth within a fairly short time (Barnosky et al., 2011). "Short" in this case is thought of as about 1 million years. The numbers of lost species are graphed versus time. Sharp peaks above the background are classified as mass extinctions.

The Earth has experienced five mass extinctions in the last 540 million years. Although the causes of those five events are not completely understood, they were clearly natural events, such as extreme temperature changes, changes in sea levels, asteroid strikes, and massive volcanic eruptions. So mass extinctions are on a completely different scale than ordinary extinctions. It is almost beyond imagination to ponder death on that level.

Of course, even in mass extinctions, the loss of some species allowed others to come into their own. The best-known mass extinction occurred when an asteroid struck Earth about 66 million years ago. Among the many species lost were the dinosaurs. However, their loss allowed mammals to blossom and eventually humans to emerge. Fortunately, humans have not had to experience such an event, at least so far.

Now some scientists believe that we are entering a sixth mass extinction, and this one is most definitely caused by human activity. Whether or not we are starting into that next event, large numbers of species are being lost, and those losses are happening at an alarming rate. But not everyone agrees that another mass extinction is underway. Some scientists refer to the IUCN Red List to support their assertion that a sixth mass extinction has not started because the number of extinctions is no greater than the background rate. However, that list does not take into account all species. It focuses mainly on birds and mammals, but shortchanges invertebrates. Unfortunately, invertebrates represent the vast majority of species. Cowie et al. (2022) examined the extinctions of mollusks and calculated that possibly 7.5%–13% of all 2 million known mollusk species have been lost since the year 1500. From those extrapolations, the rate of extinction overall is far in excess of background and suggests that we are on the cusp of a sixth mass extinction. They also found differences for extinctions among groups. Island species are faring worse than continental ones. They also note that threatened species are not treated equally. It's easy to raise concern for an endangered snow leopard, but few calls are heard to save a threatened flatworm.

Other less attractive but important species may also be affected. For example, Cavicchioli et al. (2019) examined microorganisms and climate change. We cannot see microorganisms (i.e., bacteria, archaea), but they are critical to the health of our planet. They also provide the foundation for all life on Earth. They are involved in the carbon and nutrient cycles. Many plants depend on them. Humans and animals do too. They were among the first organisms to evolve. They are the most common. They inhabit every ecosystem on Earth, even those in which they are the only living organisms. The Earth's oceans are huge, and it takes an enormous amount of energy to raise their temperature. Since about 1900, the average surface temperature has risen about 1°C (Lindsey and Dahlman, 2023). Every month in 2022 was ranked among the 10 hottest for that month. November was 1.35°C warmer than the average. At this rate, the future is not encouraging. If the warming continues, the average temperature could go up by 2.75°C or more by 2050.

So far, most of the species lost have been due to human activities, such as loss of habitat and overhunting or overfishing. However, now a more serious situation is quickly developing: the climate crisis. Scientists have been warning that we need to keep rising temperatures to less than 2°C. Already, the climate crisis is having serious effects on humans and other life on Earth. India is now the most populous country with about 1.4 billion people. Temperatures there have reached record levels. As the Earth becomes ever hotter, the situation will get worse.

The Neolithic era began 9,000 years ago, and the global mean surface temperature of the Earth varied only by +/-0.5°C. However, since the beginning of the Industrial Revolution, temperatures have changed dramatically. Nine characteristics have been used to describe the overall suitability of the Earth for life. These include climate change (e.g., CO_2 concentrations), ozone concentrations, aerosol levels, ocean acidification, and changes to freshwater and land systems. Currently, six of the nine are outside of the safe operating space for humans (Richardson et al., 2023). A complete understanding of these requires a systems approach.

Natural selection is the mechanism that powers evolution. Organisms that adapt survive and reproduce at a greater rate. The others reproduce less effectively and may die out. The warming Earth will put enormous pressure on all living organisms, including humans. So far, we have demonstrated no real interest in trying to halt or even slow this process.

Glossary

Abscission. Process by which a layer of cells dies to allow leaves or fruit to drop.

Actin. One of two main proteins in muscle tissue.

Activity-regulated protein. Protein involved in memory that might have originated from a virus.

Adherens junction. A type of connection between skin and intestinal cells.

Aerobic describes processes. Metabolic processes that occur in the presence of oxygen.

Agriculture. The process of planting, nurturing, and harvesting crops.

Alu element. A short repeated sequence of DNA.

Amino acids. Chemical building blocks of proteins. The sequence of the amino acids determines the structure and activity of the final protein.

Anaerobic describes processes. Metabolic processes that occur in the absence of oxygen.

Angiosperm. Flowering plants.

Anoikis. Process that helps to maintain the proper size and shape of organs and tissues.

Antagonistic pleiotropy theory. Theory of aging that states that a single gene may have multiple effects that manifest at different stages.

Antigen. Small molecules that elicit the production of antibodies by the adaptive immune system.

Apoptosis. Type of programmed cell death characterized by blebbing, cell shrinkage, nuclear fragmentation, chromatin condensation, chromosomal DNA fragmentation, and global mRNA decay.

ATP. Adenosine triphosphate is a molecule used to store energy.

Autophagy. Mechanism by which a cell degrades unnecessary or dysfunctional parts and recycles the material.

Bacteriophage. Virus that attacks bacteria.

Biofilm. Matrix of bacteria and other material secreted by the bacteria.

Biophilia. Concept developed by E.O. Wilson that posits that humans have a natural affinity for living things.

Bipedalism. The ability to walk on two feet.

Bystander killing. The death of cells that are lost without seeming to be infected by the human immunodeficiency virus.

Carbon cycle. The cyclic reuse of carbon in natural systems.

Carbon dioxide. A gas that is the end product of respiration in animals.

Cardiomyocytes. The contractile cells in the heart muscle that beat to pump blood.

Cartilage. Firm white connective tissue that provides structure for certain parts of the body.

Caspases. Protease enzymes that are essential in programmed cell death.

Cellulase. An enzyme that degrades cellulose.

Cellulose. A polysaccharide (sugar polymer) that is used by plants to provide support.

Chitinase. An enzyme that degrades chitin.

Chlorophyl. Pigment used in photosynthesis.

Chloroplast. Plant cell organelle that is involved in photosynthesis.

Chordata. Animal phylum that includes animals with a backbone.

Chromosome. A complex of DNA and proteins that contains the genes.

Circadian rhythm. Natural processes that follow a 24-hour pattern.

Coelom. A cavity in an animal body. Humans have a chest cavity and an abdominal cavity.

Conformation. The three-dimensional structure of proteins. It is critical to allow the function of a protein, enzyme, or antibody.

Connective tissue. Structural tissue that supports organs and cells, transports nutrients and wastes, defends against pathogens, stores fat, and repairs damaged tissues.

Cornification. Final step in the differentiation of the outer protective skin layer.

CRISPR. A method of editing genomes based on cutting DNA with the enzyme cas9 and a guide RNA.

Cucinivores. Organisms (i.e., humans) that eat primarily cooked food.

Decomposition. The breakdown of organisms after death.

Denaturation. The loss of critical three-dimensional structure of proteins. Heat, acid, and salts can denature a protein.

Dental plaque. Biofilm that develops on the teeth.

Dermis. The middle layer of the skin.

Desmosome. A type of junction between skin and intestinal cells.

Disposable soma theory. Theory of aging that states that organisms have many requirements that require compromises.

Division. Largest classification of plants.

DNA damage theory. Theory of aging that states that aging is due to an accumulation of damaged genes.

Duchenne's smile. A true smile that reflects a person emotions.

Ectoderm. The outermost layer of tissues in an animal.

Elastin. An elastic protein that can be stretched. It is a major component of lung tissue, for example.

Emulsion. A mixture of small droplets of one liquid in another that are not miscible.

Endoderm. The innermost layer of tissues in an animal.

Endosperm. Part of a seed that provides a food supply for the embryo.

Endosymbiosis. A symbiotic relationship in which an organism lives inside another. One example is the mitochondria that live in all of our cells.

Enzyme. Protein that catalyzes a biochemical reaction.

Epidermis. The outermost layer of the skin.

Epigenetics. Heritable changes that do not involve the DNA sequence (e.g., DNA methylation, histone compositions).

Erythrocytes. The red blood cells that carry oxygen in the blood.

Eukaryote. Cells that have a nucleus and other organelles surrounded by a membrane and organized chromosomes.

Excitotoxicity. Degradation of neurons from overuse.

Exoskeleton. An external hard structure that provides support for many invertebrates.

Fatty acid. An organic acid that is used to form triglycerides.

Fecal transplant. The use of fecal pellets to reseed the intestinal biome.

Fossilization. The process of a plant or animal become petrified.

Fruiting body. The reproductive organ of fungi. The best example is the mushrooms that we see on a pizza.

Gametophyte. The gamete-producing, haploid phase of plant reproduction.

Germination. The activation of a seed or spore to begin to grow.

Glycerol. Odorless liquid that provides the backbone of triglycerides, the most common energy-storage compound.

Gymnosperm. Nonflowering seed plants, such as pines, spruces, and firs.

Heliobacter pylori. A bacterium that is associated with stomach ulcers.

Hemicellulase. Enzyme that breaks down hemicellulose.

Hemocyanin. A copper-based blood component that carries oxygen.

Hemoglobin. The protein complex that carries the iron molecules that bind oxygen in red blood cells.

Human Biome Project. Project that cataloged all of the organisms living in and on a human.

Human Genome Project. Project that sequenced the human genome.

Humus. The dark, organic matter that is found in soil.

Hunter-gatherer. A stage in the development of humans that predates agriculture.

Hydrogen bonds. Weak electrostatic attraction between a hydrogen and another positively charged ion.

Hydrophilic. Having a tendency to dissolve in water.

Hydrophobic. Having a tendency to repel water.

Hyphae. Filaments that make up the mycelium of fungi.

Hypodermis. Bottom of three layers of the skin.

Invertebrate. An organism lacking a backbone.

Keratin. The primary protein that makes up hair.

Lactose. The sugar found in milk. It is a disaccharide of galactose and glucose.

Lactose intolerant. The inability to digest lactose.

Latent. Existing in a dormant form.

Ligament. Connective tissue that joins bones.

Lignin. A high-molecular-mass polymer that makes up plant cell walls.

Mass extinction. A sudden reduction in the number of species and organisms living on Earth. While extinctions happen continually, these events involve the loss of the great majority of living organisms on Earth.

Melanin. A pigment that gives dark color to the skin, hair, and eyes.

Melanoma. A dangerous type of cancer that originates in skin cells called melanocytes.

Mesoderm. The middle layer of three tissue layers in most animals.

Metamorphosis. Process in which an animal changes dramatically as it grows and differentiates.

Methane. A colorless, odorless gas that is flammable and has a chemical structure of CH_4.

Methanogenic organism. Microorganisms that produce methane as a metabolic by-product in hypoxic conditions.

Methemoglobin. Another form of hemoglobin, in which the iron is in the +3 state.

Microbiome. The collection of microorganisms that live in and on the human body.

MicroRNA. Small, noncoding RNAs that regulate translation.

Midden. A refuse heap.

Mitochondria. Cell organelle that is involved in the production of energy and in programmed cell death.

Mucus membrane. The lining of the mouth, nasal passages, and genital tissues.

Mutation accumulation theory. Theory of aging that states that genes that are deleterious later in life have little negative selection.

Mycelium. A network of fungal threads or hyphae.

Myoglobin. An iron-containing protein in muscle.

Myosin. One of two main proteins that make up muscle tissue.

NAD. Nicotine adenine diphosphate is a molecule that is involved in energy production and use in living organisms.

Natural selection. The concept that states that organism that survive better are able to reproduce more successfully. It is the driving force behind evolution.

Neandertals. Species or subspecies of humans in Eurasia that died out about 40,000 years ago. However, they did interbreed with modern humans.

Necroptosis. Type of necrosis or programmed cell death.

Necrosis. Death of cells in a relatively unorganized manner.

Neuron. Nerve cell that is responsible for the long-range communication of nerve impulses.

Nitrogen fixation. The ability to transform elemental nitrogen into usable compounds.

Non-REM sleep. Phase of sleep in which there is no rapid eye movement. Fewer dreams occur during this phase.

Omnivore. A living organism that consumes both animal and plant material.

Osteoarthritis. A disorder characterized by deterioration of the joints.

Oxidation. The process of adding an oxygen to a compound.

Oxidizing atmosphere. An atmosphere that contains oxygen, such as the current atmosphere of the Earth.

Paleovirus. An ancient virus that is now extinct.

Pandemic. An epidemic that involves the entire world.

Pattern-recognition receptors. Receptors that recognize common markers on pathogens.

Petrification. Process in which the organic material in a plant is replaced by minerals or the spaces in the plant are filled with minerals. That is, the process by which organic matter is turned to stone.

Pheromone. A chemical substance released by an animal as a means of signaling.

Phloem. Plant tissue that carries food from the leaves to the rest of the plant.

Photosynthesis. The process by which plants capture the energy of the sun and transform it into chemical energy.

Phylum. The highest level of classification of animals.

Plasma. The liquid portion of the blood when the cells have been removed.

Plasmid. A small circular DNA found in bacteria.

Pond scum. A collection of micro- and macroscopic plants and animals found in small bodies of water.

Programmed cell death. Highly ordered process that result in the death of a cell. The triggering event can be internal or external. Autophagy and apoptosis are two types of programmed cell death.

Prokaryote. A cell lacking a nucleus and other organelles that is surrounded by a membrane. These include bacteria and blue-green algae.

Protein. A polymer of amino acids that form enzymes, structural elements, and antibodies.

Proteostasis. The process of maintaining the correct types and levels of proteins in a cell. It involves both production and destruction of proteins.

Protista. Any eukaryote that is not an animal, plant, or fungus.

Pyroptosis. A type of programmed cell death that involves a distinct set of caspases and results in a highly inflammatory reaction.

Quorum sensing. The ability of bacteria to communicate with each other by exchanging diffusible signaling molecules to determine whether they must reduce the number of individuals in the colony to ensure long-term survival of the colony.

Reactive oxygen species. Compounds that readily react with biological molecules; they include peroxides, superoxides, and hydroxyl radicals.

Red queen. Theory in which one organism has to continually evolve new defenses to keep up with its predators.

Redox reaction. A chemical reaction that involves the transfer of electrons between reactants.

Reducing atmosphere. Atmosphere that lacks oxygen. The first atmosphere on primitive Earth was a reducing atmosphere.

Reliability theory. Theory of aging that state that redundant systems fail over time.

REM sleep. Sleep phase characterized by rapid eye movement and dreaming.

Retrovirus. A virus that uses RNA for its genetic material, but uses reverse transcription to make a DNA copy during replication.

Rhizome. A plant stem that grows underground as a form of replication.

Root cap. Structure that protects the growing tip of the root.

Sediment. Material that deposits in layers over time.

Segmentation. Repetition of similar tissues along the axis of animals with bilateral symmetry.

Soil horizon. A layer in the soil.

Sporophyte. The diploid and asexual phase in plant reproduction.

Surfactant. Chemical compound that reduces the surface tension between two liquids or a liquid and gas.

Symbiosis. A mutually beneficial relationship between organisms.

Symmetry. In living organisms, symmetry refers to the ability to cut the organism into two mirror images.

Telomers. Structures at the end of chromosomes that are related to aging.

Tendon. Connective tissue that connects a muscle to a bone.

Thermoregulation. The ability of an organism to regulate the temperature of its own body.

Tight junction. A type of connection between skin and intestinal cells.

Topsoil. The upper layer of soil that tends to contain high levels of organic matter.

Transcription factor. Protein that regulates the genes.

Ubiquitin-proteasome system. Mechanism in cells in which proteins to be degraded are labeled with ubiquitin and then degraded in the proteasome.

Vascular plant. Plants that contain systems for transporting food and water up and down stems.

Vertebrate. An animal that possesses a backbone.

Wear and tear theory. Theory of aging that states that, like machines, our bodies simply wear out.

Xylem. A plant tissue that transports water.

Zooplankton. Microscopic animals.

References

Abelson MB, Ousler G, Shapiro A, et al. (2016) The form and function of meibomian glands. *Review of Ophthalmology*. Retrieved from: https://www.reviewofophthalmology.com/article/the-form-and-function-of-meibomian-glands. November 2, 2023.

Abe M, Kuroda R (2019) The development of CRISPR for a mollusc establishes the formin *Lsdia1* as the long-sought gene for snail dextral/sinistral coiling. *Development* 146(9): dev175976.

Ackermann M, Chao L, Bergstron CT, et al. (2007) On the evolutionary origin of aging. *Aging Cell* 6: 235–244.

Adolph KE, Berger SE, Leo AJ (2011) Developmental continuity? Crawling, cruising, and walking. *Developmental Science* 14: 306–318.

Ahmad HI, Ahmad MJ, Jabbir F, et al. (2020) The domestication makeup: evolution, survival, and challenges. *Frontiers in Ecology and Evolution* 8: 1–17.

Alexander RM (2004) Bipedal animals, and their differences from humans. *Journal of Anatomy* 204: 321–330.

Alkire MT, Hudetz AG, Tononi G (2008) Consciousness and anesthesia. *Science* 322: 876–880.

Allen KD, Thoma LM, Golightly YM (2022) Epidemiology of osteoarthritis. *Osteoarthritis and Cartilage* 30: 184–195.

Allen RP (2015) Restless leg syndrome/Willis-Ekbom disease pathophysiology. *Sleep Medicine Clinics* 10: 207–214.

Allocati N, Masulli M, Di Ilio C, et al. (2015) Die for the community: an overview of programmed cell death in bacteria. *Cell Death and Disease* 6: e1609.

Almécija S, Smaers JB, Jungers WL (2015) The evolution of human and ape hand proportions. *Nature Communications* 6: 7717.

Aman Y, Schmauck-Medina T, Hansen M, et al. (2021) Autophagy in healthy aging and disease. *Nature Aging* 1: 634–650.

Amarante-Mendes GP, Adjemian S, Branco LM, et al. (2018) Pattern recognition receptors and the host cell death molecular machinery. *Frontiers in Immunology* 9: 2379.

Ambros V (2011) MicroRNAs and developmental timing. *Current Opinion in Genetics and Development* 21: 511–517.

Ameisen J (2002) On the origin, evolution, and nature of programmed cell death: a timeline of four billion years. *Cell Death Differentiation* 9: 367–393.

AMNH (nd) *Trilobites in history*. American Museum of Natural History. Retrieved from: https://www.amnh.org/research/paleontology/collections/fossil-invertebrate-collection/trilobite-website/the-trilobite-files/trilobites-in-history. November 1, 2023.

Amodio P, Josef N, Shashar N, et al. (2021) Bipedal locomotion in *Octopus vulgaris*: a complementary observation and some preliminary considerations. *Ecology and Evolution* 11: 3679–3684.

Anafi RC, Kayser MS, Raizen DM (2019) Exploring phylogeny to find the function of sleep. *Nature Reviews Neuroscience* 20: 109–116.

Anderson JL (2020) The musk ox and me. *New Yorker*, August 10, 2020.

Andrewes WJH (2006) A chronicle of timekeeping. *Scientific American Special Edition* 16: 46–55.

Antcliff A, McCullough LD, Tsvetkov AS (2021) G-Quadruplexes and the DNA/RNA helicase DHX36 in health, disease, and aging. *Aging (Albany, NY)* 13: 25578–25587.

Araki S, Shirahata A (2020) Vitamin K deficiency bleeding in infancy. *Nutrients* 12(3): 780.

Arbib MA, Liebal K, Pika S (2008) Primate vocalization, gesture, and the evolution of human language. *Current Anthropology* 49: 1053–1076.

Armanios M, Blackburn EH (2012) The telomere syndromes. *Nature Reviews Genetics* 13: 693–704.

Asad H, Carpenter DO (2018) Effects of climate change on the spread of zika virus: a public health threat. *Reviews on Environmental Health* 33: 31–42.

Ashley J, Cordy B, Lucia D, et al. (2018) Retrovirus-like Gag protein Arc1 binds RNA and traffics across synaptic boutons. *Cell* 172: 262–274.Asplund J, Wardle DA (2017) How lichens impact on terrestrial community and ecosystem properties. *Biological Reviews* 92: 1720–1738.

Aubert G, Lansdorp PM (2008) Telomeres and aging. *Physiological Review* 88: 557–579.

Ayyar VS, Sukumaran S (2021) Circadian rhythms: influence on physiology, pharmacology, and therapeutic interventions. *Journal of Pharmacokinetics Pharmacodynamics* 48: 321–338.

Baar MP, Brandt RMC, Putavet DA, et al. (2017) Targeted apoptosis of senescent cells restores tissue homeostasis in response to chemotoxicity and aging. *Cell* 169: 132–147.

Baehrecke EH (2003) Autophagic programmed cell death in *Drosophila*. *Cell Death and Differentiation* 10: 940–945.

Bais HP, Park S-W, Weir TL, et al. (2004) How plants communicate using the underground information superhighway. *Trends in Plant Science* 9: 26–32.

Ballabio A (2016) The awesome lysosome. *EMBO Molecular Medicine* 8: 73–76.

Bapat SP, Whitty C, Mowery CT, et al. (2022) Obesity alters pathology and treatment response in inflammatory disease. *Nature* 604: 337–342.

Barnes BA, Raff MC (1999) Axonal control of oligodendrocyte development. *Journal of Cell Biology* 147: 1123–1128.

Barnosky AD, Matzke N, Tomiya S, et al. (2011) Has the Earth's sixth mass extinction already arrived? *Nature* 471: 51–57.

Barud NR, Pollard KS (2020) Population genetics in the human microbiome. *Trends in Genetics* 36: 53–67.

Basit H, Tariq MA, Siccardi MA (2021) Anatomy, head and neck, mastication muscles. In: *StatPearls*. Retrieved from: https://www.ncbi.nlm.nih.gov/books/NBK541027/. November 2, 2023.

Battin TJ, Luyssaert S, Kaplan LA, et al. (2009) The boundless carbon cycle. *Nature Geoscience* 2: 598–600.

Bauer U, Müller UK, Poppinga S (2021) Complexity and diversity of motion amplification and control strategies in motile carnivorous plant traps. *Proceedings of the Royal Society B* 288: 20210771.

Bayles KW (2007) The biological role of death and lysis in biofilm development. *Nature Reviews Microbiology* 5: 721–726.

Becker B, Marin B (2009) Streptophyte algae and the origin of embryophytes. *Annals of Botany* 103: 999–1004.

Becraft ED, Lau Vetter MCY, Bezuidt OKI, et al. (2021) Evolutionary stasis of a deep subsurface microbial lineage. *ISME Journal* 15: 2830–2842.

Beecher MD (2021) Why are no animal communication systems simple languages? *Frontiers in Psychology* 12: 602635. https://doi.org/10.3389/fpsyg.2021.602635.

Beekman LM, Ratnieks FLW (2000) Long-range foraging by the honey-bee, *Apis mellifera*. *Functional Ecology* 14: 490–496.

Beijerinck MW (1901) Über oligonitrophile Mikroben. *Zentralblatt für Bakteriologie, Parasitenkunde, Infektionskrankheiten und Hygiene. Abteilung II* 7: 561–582.

Belongia EA, Naleway AL (2003) Smallpox vaccine: the good, the bad, and the ugly. *Clinical Medicine and Research* 1: 87–92.

Berg C, Engels I, Rothbart A (2001) Human mature red blood cells express caspase-3 and caspase-8, but are devoid of mitochondrial regulators of apoptosis. *Cell Death Differentiation* 8: 1197–1206.

Berghe TV, Vanlangenakker N, Parthoens E, et al. (2010) Necroptosis, necrosis and secondary necrosis converge on similar cellular disintegration features. *Cell Death and Differentiation* 17: 922–930.

Berut A, Chauvet H, Legue V, et al. (2018) Gravisensors in plant cells behave like an active granular liquid. *Proceedings of the National Academy of Sciences USA* 115: 5123–5128.

Bibell G (2008) *What people have said about Linnaeus*. Uppsala University. Retrieved from: https://web.archive.org/web/20110513033923/http://www.linnaeus.uu.se/online/life/8_3.html. November 1, 2023.

Black B (2013) Of barosaurus and blood pressure. *National Geographic*. Retrieved from: https://www.nationalgeographic.com/science/article/of-barosaurus-and-blood-pressure. November 2, 2023.

Blackledge TA, Scharff N, Coddington JA, et al. (2009) Reconstructing web evolution and spider diversification in the molecular era. *Proceedings of the National Academy of Sciences USA* 106: 5229–5234.

Blanchard C (2017) A history into genetic and epigenetic evolution of food tolerance: how humanity rapidly evolved by drinking milk and eating wheat. *Current Opinion in Allergy and Clinical Immunology* 17: 460–464.

Blaser ML (2005) An endangered species in the stomach? *Scientific American* 292: 38–45.

Boguszewska K, Szewczuk M, Kaźmierczak-Barańska J, et al. (2020) The similarities between human mitochondria and bacteria in the context of structure, genome, and base excision repair system. *Molecules* 25: 2857.

Bohn LE (2009) Q&A: "Lucy" discoverer Donald C. Johanson (March 04, 2009). *Time Magazine*. Retrieved from: https://content.time.com/time/health/article/0,8599,1882969,00.html. June 30, 2023.

Boskey AL, Coleman R (2010) Aging and bone. *Journal of Dental Research* 89: 1333–1348.

Bowen R (nd) *Physiology of Vomiting*. Colorado State University. Retrieved from: http://www.vivo.colostate.edu/hbooks/pathphys/digestion/stomach/vomiting.html. November 2, 2023.

Brandner JM, Zorn-Kruppa M, Yoshida T, et al. (2015) Epidermal tight junctions in health and disease. *Tissue Barriers* 3(1-2): e974451.

Brizuela M, Winters R (2023) Histology, oral mucosa. [Updated May 8, 2023]. *StatPearls [Internet]*. StatPearls Publishing. Retrieved from: https://www.ncbi.nlm.nih.gov/books/NBK572115/. November 8, 2023.

Brokaw A, Furuta A, Dacanay M, et al. (2021) Bacterial and host determinants of group B streptococcal vaginal colonization and ascending infection in pregnancy. *Frontiers in Cellular and Infection Microbiology* 11. https://doi.org/10.3389/fcimb.2021.720789.

Broly P, Deville P, Maillet S (2013) The origin of terrestrial isopods (Crustacea: Isopoda: Oniscidea). *Ecology and Evolution* 27: 461–476.

Brooks RC, Garratt MG (2017) Life history evolution, reproduction, and the origins of sex-dependent aging and longevity. *Annuals of the New York Academy of Science* 1389: 92–107.

Brown GD, Denning DW, Levitz SM (2012) Tackling human fungal infections. *Science* 336: 647.

Brown GG, Barois I, Lavelle P (2000) Regulation of soil organic matter dynamics and microbial activity in the drilosphere and the role of interactions with other edaphic functional domains. *European Journal of Soil Biology* 36: 177–198.

Brown M, Goldstein J (1976) Receptor-mediated control of cholesterol metabolism. *Science* 191: 150–154.

Brown MW, Spiegel FW, Silberman JD (2009) Phylogeny of the "forgotten" cellular slime mold, *Fonticula alba*, reveals a key evolutionary branch within Opisthokonta. *Molecular Biology and Evolution* 26: 2699–2709.

Bruneau B (2009) Benoit Bruneau talks about the evolution of the four chambers of the heart from frogs to mammals. *National Science Foundation.* News Release 09-164. Retrieved from: https://www.nsf.gov/news/news_videos.jsp?cntn_id=115520&media_id= 65577&org=NSF. November 2, 2023.

Burgess IF (2004) Human lice and their control. *Annual Review of Entomology* 49: 457–481.

Burmester T (2001) Molecular evolution of the arthropod hemocyanin superfamily. *Molecular Biology and Evolution* 18: 184–195.

Butenandt A, Beckmann R, Hecker E (1961) Über den sexuallockstoff des seidenspinners. 1. Der biologische test und die isolierung des reinen sexuallockstoffes bombykol. *Hoppe-Seyler's Zeitschrift für Physiologische Chemie* 324: 71–83.

Buts L, Lah J, Dao-Thi MH, et al. (2005) Toxin-antitoxin modules as bacterial metabolic stress managers. *Trends Biochemical Science* 30: 672–679.

Byrd A, Belkaid Y, Segre J (2018) The human skin microbiome. *Nature Reviews Microbiology* 16: 143–155.

Campbell SS, Tobler I (1984) Animal sleep: a review of sleep duration across phylogeny. *Neuroscience and Biobehavioral Reviews* 8: 269–300.

Cannon MJ, Schmid DS, Hyde TB (2010) Review of cytomegalovirus seroprevalence and demographic characteristics associated with infection. *Reviews in Medical Virology* 20: 202–213.

Cantlon JF, Brannon EM, Carter EJ, et al. (2006) Functional imaging of numerical processing in adults and 4-y-old children. *Public Library of Science Biology* 4(5): e125.

Capone KA, Dowd SE, Stamatas GN, et al. (2011) Diversity of the human skin microbiome early in life. *Journal of Investigative Dermatology* 131: 2026–2032.

Cardoso-Moreira M, Halbert J, Valloton D, et al. (2019) Gene expression across mammalian organ development. *Nature* 571: 505–509.

Carroll LC (1909). *Through the Looking Glass: And What Alice Found There.* Dodge Publishing Company.

Carskadon MA, Dement WC (2011) Monitoring and staging human sleep. In: *Principles and Practice of Sleep Medicine,* 5th edition (Kryger MH, Roth T, Dement WC, eds). Elsevier Saunders, pp. 16–26.

Casselman A (2007) Strange but true: the largest organism on Earth is a fungus. *Scientific American.* Retrieved from: https://www.scientificamerican.com/article/strange-but-true-largest-organism-is-fungus/. November 1, 2023.

Cavicchioli R, Ripple WJ, Timmis KN, et al. (2019) Scientists' warning to humanity: microorganisms and climate change. *Nature Reviews Microbiology* 17: 569–586. https://www.nature.com/articles/s41579-019-0222-5#citeas.

Chakravarti D, LaBella KA, DePinho RA (2021) Telomeres: history, health, and hallmarks of aging. *Cell* 184: 306–322.

Chen F, Liu X, Yu C, et al. (2017) Water lilies as emerging models for Darwin's abominable mystery. *Horticulture Research* 4: 17051.

Chiu ML, Goulet DR, Teplyakov A, et al. (2019) Antibody structure and function: the basis for engineering therapeutics. *Antibodies (Basel)* 8: 55.

Choi BC, Hunter DJ, Tsou W, et al. (2005) Diseases of comfort: primary cause of death in the 22nd century. *Journal of Epidemiology Community Health* 59: 1030–1034.

Christenhusz MJM, Byng JW (2016) The number of known plants species in the world and its annual increase. *Phytotaxa* 261: 201–217.

Chu DM, Ma J, Prince AL, et al. (2017) Maturation of the infant microbiome community structure and function across multiple body sites and in relation to mode of delivery. *Nature Medicine* 23: 314–326.

Clarke B (2008) Normal bone anatomy and physiology. *Clinical Journal of the American Society of Nephrology* 3(Suppl 3): S131–S139.

Clegg RJ, Dyson RJ, Kreft J-U (2014) Repair rather than segregation of damage is the optimal unicellular aging strategy. *BioMed Central Biology* 12: 52.

Cliffe LJ, Humphreys NE, Lane TE, et al. (2005) Accelerated intestinal epithelial cell turnover: a new mechanism of parasite expulsion. *Science* 308: 1463–1465.

Cobb BA (1914) Nematodes and their relationships. *Yearbook of the Department of Agriculture.* p. 472. Retrieved from: https://web.archive.org/web/20160326113323/http://naldc.nal.usda.gov/naldc/download.xhtml?id=IND43748196&content=PDF. November 1, 2023.

Cohen FS (2016) How viruses invade cells. *Biophysics Journal* 110: 1028–1032.

Cohen JI (2020) Herpesvirus latency. *Journal of Clinical Investigation* 130: 3361–3369.

Coleman E, Inusa B (2007) Sickle cell anemia: targeting the role of fetal hemoglobin in therapy. *Clinical Pediatrics (Philadelphia)* 46: 386–391.

Collins JJ, III (2017) Platyhelminthes. *Current Biology* 27: R252–R256.

Collins JT, Nguyen A, Badireddy M (2022) Anatomy, abdomen and pelvis, small intestine. *StatPearls.* StatPearls Publishing. Retrieved from: https://www.ncbi.nlm.nih.gov/books/NBK459366/. November 2, 2023.

Condamine FL, Silverstroc D, Koppelhus EB, et al. (2020) The rise of angiosperms pushed conifers to decline during global cooling. *Proceedings of the National Academy of Sciences USA* 117: 28867–28875.

Condon JC, Jeyasuria P, Faust JM, et al. (2004) Surfactant protein secreted by the maturing mouse fetal lung acts as a hormone that signals the initiation of parturition. *Proceedings of the National Academy of Sciences USA* 101: 4978–4983.

Conover MR (2019) Numbers of human fatalities, injuries, and illnesses in the United States due to wildlife. *Human–Wildlife Interactions* 13: 264–276.

Cordero GA (2017) The turtle's shell. *Current Biology* 27: PR168–R169.

Corniani G, Saal HP (2020) Tactile innervation densities across the whole body. *Journal of Neurophysiology* 124: 1229–1240.

Costanzo A, Fausti F, Spallone G, Moretti F, Narcisi A, Botti E (2015) Programmed cell death in the skin. *International Journal of Developmental Biology* 59: 73–78.

Counter P, Wilson JA (2004) The management of simple snoring. *Sleep Medicine Reviews* 8: 433–441.

Cousteau J-Y (1981) Ocean policy and reasonable utopias. *Forum* 16, no. 5(Summer): 900.

Cowie RH, Bouchet P, Fontaine B (2022) The sixth mass extinction: fact, fiction or speculation? *Biological Reviews* 97: 640–663.

Creux N, Harmer S (2019) Circadian rhythms in plants. *Cold Spring Harbor Perspectives in Biology* 11: a034611.

Dąbrowska AK, Spano F, Derler S, et al. (2018) The relationship between skin function, barrier properties, and body-dependent factors. *Skin Research and Technology* 24: 165–174.

Damen WGM, Saridaki T, Averof M (2002) A common evolutionary origin for wings, breathing organs, and spinnerets. *Current Biology* 12: 1711–1716.

Daneman R, Prat A (2015) The blood-brain barrier. *Cold Spring Harbor Perspectives in Biology* 7(1): a020412.

Daniels CB, Orgeig S, Sullivan LC, et al. (2004) The origin and evolution of the surfactant system in fish: insights into the evolution of lungs and swim bladders. *Physiology and Biochemical Zoology* 77: 732–749.

Darwin C (1880) *The Power of Movement in Plants.* D. Appleton.

Daugas E, Candé C, Kroemer G (2001) Erythrocytes: death of a mummy. *Cell Death Differentiation* 8: 1131–1133.

Dean L (2005a) The ABO blood group. In: *Blood Groups and Red Cell Antigens.* National Center for Biotechnology Information, Bethesda, p. 31.

Dean L (2005b) The Rh blood group. In: *Blood Groups and Red Cell Antigens.* National Center for Biotechnology Information, Bethesda, p. 49.

Deaglio S, Amoroso A, Rinaldi M, Boffini M (2020) HLA typing in lung transplantation: Does high resolution fit all? *Annals of Translational Medicine* 8: 45.

De Groof TWM, Elder EG, Lim EY, et al. (2021) Targeting the latent human cytomegalovirus reservoir for T-cell-mediated killing with virus-specific nanobodies. *Nature Communications* 12: 4436.

Delage B, Angelo G, Drake VJ, Higdon J (2019) *Essential Fatty Acids.* Linus Pauling Institute. Retrieved from: https://lpi.oregonstate.edu/mic/other-nutrients/essential-fatty-acids#authors-reviewers. November 2, 2023.

Delgado-Baquerizo M, Oliverio AM, Brewer TE, et al. (2018) A global atlas of the dominant bacteria found in soil. *Science* 359: 320–325.

DeMaria S, Ngai J (2010) The cell biology of smell. *Journal of Cell Biology* 191: 443–452.

Dempsey JA, Veasey SC, Morgan BJ, et al. (2010) Pathophysiology of sleep apnea. *Physiological Reviews* 90: 47–112.

Depicolzuane L, Phelps DS, Floros J (2022) Surfactant protein-A function: knowledge gained from SP-A knockout mice. *Frontiers in Pediatrics* 9: 799693 https://doi.org/10.3389/fped.2021.799693.

Desseilles M, Dang-Vu TT, Sterpenich V, et al. (2011) Cognitive and emotional processes during dreaming: a neuroimaging view. *Consciousness and Cognition* 20: 998–1008.

de Vries J, Archibald JM (2018) Plant evolution: landmarks on the path to terrestrial life. *New Phytologist* 217: 1428–1434.

Diamond J (2002) Evolution, consequences and future of plant and animal domestication. *Nature* 418: 700–707.

Dieleman JL, Cao J, Chapin A, et al. (2020) US health care spending by payer and health condition, 1996–2016. *Journal of the American Medical Association* 323: 863–884.

Diekelmann S, Born J (2010) The memory function of sleep. *Nature Reviews Neuroscience* 11: 114–126.

Dimitrov DS (2004) Virus entry: molecular mechanisms and biomedical applications. *Nature Reviews Microbiology* 2: 109–122.

Damen WGM, Saridaki T, Averof Ml (2002) Diverse adaptations of an ancestral gill: A common evolutionary origin for wings, breathing organs, and spinnerets. *Current Biology* 12: 1711–1716.

Dixon R, Kahn D (2004) Genetic regulation of biological nitrogen fixation. *Nature Reviews Microbiology* 2: 621–631.

Doitsh G, Galloway NL, Geng X, et al. (2014) Cell death by pyroptosis drives CD4 T-cell depletion in HIV-1 infection. [published correction appears in *Nature* (2017) 544(7648): 124]. *Nature* 505: 509–514.

Domínguez-Rodrigo M, Bunn HT, Mabulla AZP, et al. (2014) On meat eating and human evolution: a taphonomic analysis of BK4b (Upper Bed II, Olduvai Gorge, Tanzania), and its bearing on hominin megafaunal consumption. *Quaternary International* 322–323: 129–152.

Donoghue P, Paps J (2020) Plant evolution: assembling land plants. *Current Biology* 30: R81–R83.

Dreischer P, Duszenko M, Stein J, et al. (2022) Eryptosis: programmed death of nucleus-free, iron-filled blood cells. *Cells* 11: 503.

Dunne J, Mercuri AM, Evershed RP, et al. (2016) Earliest direct evidence of plant processing in prehistoric Saharan pottery. *Nature Plant* 3: 16194.

Dunn R (2011) The top ten deadliest animals of our evolutionary past. *Smithsonian Magazine.* Retrieved from: https://www.smithsonianmag.com/science-nature/the-top-ten-deadliest-animals-of-our-evolutionary-past-18257965/. November 2, 2023.

Durack J, Kimes NE, Lin DL, et al. (2018) Delayed gut microbiota development in high-risk for asthma infants is temporarily modifiable by *Lactobacillus* supplementation. *Nature Communications* 9: 707.

Dyble M, Thorley J, Page AE, et al. (2019) Engagement in agricultural work is associated with reduced leisure time among Agta hunter-gatherers. *Nature Human Behavior* 3: 792–796.

Dyson HJ, Wright PE (2005) Intrinsically unstructured proteins and their function. *Nature Reviews Molecular Cell Biology* 6: 197–208

Eberle U (2022) The science behind your cheese. *Smithsonian Magazine*. Retrieved from: https://www.smithsonianmag.com/science-nature/the-science-behind-your-cheese-180981199/. November 1, 2023.

Eckhart L, Lippens S, Tschachler E, et al. (2013) Cell death by cornification. *Biochimica et Biophysica Acta* 1833: 3471–3480.

Eiberg H, Troelsen J, Nielsen M, et al. (2008) Blue eye color in humans may be caused by a perfectly associated founder mutation in a regulatory element located within the *HERC2* gene inhibiting *OCA2* expression. *Human Genetics* 123: 177–187.

Eiseley L (1953) The flow of the river. *American Scholar* 22: 451–458.

Ekman P, Friesen WV (1982) Felt, false, and miserable smiles. *Journal of Nonverbal Behavior* 6: 238–252.

Ekman P, Friesen WV, Davidson RJ (1990) The Duchenne smile: emotional expression and brain physiology. *Journal of Personality and Social Psychology* 58: 342–353.

El Nahas AM, Bello AK (2005) Chronic kidney disease: the global challenge. *Lancet* 365: 331–340.

Emerman M, Malik HS (2010) Paleovirology—modern consequences of ancient viruses. *Public Library of Science Biology* 8(2): e1000301.

Epstein G, Smale DA (2017) *Undaria pinnatifida*: a case study to highlight challenges in marine invasion ecology and management. *Ecology and Evolution* 7: 8624–8642.

Erard M (2009) How many languages? Linguists discover new tongues in China. *Science* 324: 332–333.

Erekat NS (2022) Apoptosis and its therapeutic implications in neurodegenerative diseases. *Clinical Anatomy* 35: 65–78.

Eriksson N, Macpherson JM, Tung JY, et al. (2010) Web-based, participant-driven studies yield novel genetic associations for common traits. *Public Library of Science Genetics* 6(6): e1000993.

Exploratorium (2023) Science of eggs. *The Exploratorium*. Retrieved from: https://www.exploratorium.edu/explore/cooking/egg-science#:~:text=Egg%20yolk%20contains%20a%20number,amino%20acids%20that%20attract%20water. November 1, 2023.

Fabré J-H (1914) *Social Life in the Insect World* (translated by B. Miall). The Century Co.

Farji-Brener AG, Werenkraut V (2017) The effects of ant nests on soil fertility and plant performance: a meta-analysis. *Journal of Animal Ecology* 86: 866–877.

Feder HM, Johnson BJB, O'Connell S, et al., Ad Hoc International Lyme Disease Group (2007) A critical appraisal of "chronic Lyme disease." *New England Journal of Medicine* 357: 1422–1430.

Feig C, Peter ME (2007) How apoptosis got the immune system in shape. *European Journal of Immunology* 37: S61–S70.

Feng X (2009) Chemical and biochemical basis of cell-bone matrix interaction in health and disease. *Currents in Chemical Biology* 3: 189–196.

Ference BA, Kastelein JJP, Catapano AL (2020) Lipids and lipoproteins in 2020. *Journal of the American Medical Association* 324: 595–596.

Ferus M, Pietrucci F, Saitta AM, et al. (2017) Formation of nucleobases in a Mille-Urey reducing atmosphere. *Proceedings of the National Academy of Sciences USA* 114: 4306–4311.

Finch JT, Perutz MF, Bertle JF, et al. (1973) Structure of sickled erythrocytes and of sickle-cell hemoglobin fibers (electron microscopy/x-ray diffraction/molecular arrangement/helices). *Proceedings of the National Academy of Sciences USA* 70: 718–722.

Fitch WT (2018) The biology and evolution of speech: a comparative analysis. *Annual Review of Linguistics* 4: 255–279.

Flemming HC, Wuertz S (2019) Bacteria and archaea on Earth and their abundance in biofilms. *Nature Reviews Microbiology* 17: 247–260.

Fletcher DA, Mullins RD (2010) Cell mechanics and the cytoskeleton. *Nature* 463: 485–492.Flinker A, Lorzeniewsk A, Shestyuk AY, et al. (2015) Redefining the role of Broca's area in speech. *Proceedings of the National Academy of Sciences USA* 112: 2871–2875.

Föller M, Geiger C, Mahmud H, Nicolay J, Lang F (2008) Stimulation of suicidal erythrocyte death by amantadine. *European Journal of Pharmacology* 581: 13–18.

Follmann H, Brownson C (2009) Darwin's warm little pond revisited: from molecules to the origin of life. *Naturwissenschaften* 96: 1265–1292.

Forrest LL, Davis EC, Long DG, Crandall-Stotler BJ, Clark A, Hollingsworth ML (2006) Unraveling the evolutionary history of the liverworts (Marchantiophyta): multiple taxa, genomes and analyses. *Bryologist* 109: 303–334.

Forsgård RA (2019) Lactose digestion in humans: intestinal lactase appears to be constitutive whereas the colonic microbiome is adaptable. *American Journal of Clinical Nutrition* 110: 273–279.

Fortuna A (2012) The soil biota. *Nature Education Knowledge* 3(10): 1.

Foster RG, Roenneberg T (2008) Human responses to the geophysical daily, annual and lunar cycles. *Current Biology* 18: R784–R794.

Frąc M, Hannula SE, Bełka M, Jędryczka M (2018) Fungal biodiversity and their role in soil health. *Frontiers in Microbiology* 9: 707. https://doi.org/10.3389/fmicb.2018.00707.

Franchini M, Mannucci PM (2012) Past, present and future of hemophilia: a narrative review. *Orphanet Journal of Rare Diseases* 7: 24.

Franklin B (1774) *Philosophical Transactions of the Royal Society* 64(Part 1). doi.org/10.1098/rstl.1774.0044

Franklinos LHV, Jones KE, Redding DW, et al. (2019) The effect of global change on mosquito-borne disease. *Lancet Infectious Diseases* 19: e302–e312.

Frantz L, Bradley D, Larson G, et al. (2020) Animal domestication in the era of ancient genomics. *Nature Reviews Genetics* 21: 449–460.

Frazer J (2013) Nematode roundworms own this place. *Scientific American.* Retrieved from: https://blogs.scientificamerican.com/artful-amoeba/parasitic-roundworms-own-this-place/. November 1, 2023.

Freiberg AS (2020) Why we sleep: a hypothesis for an ultimate or evolutionary origin for sleep and other physiological rhythms. *Journal of Circadian Rhythms* 18: 2.

Fricker M, Tolkovsky AM, Borutaite V, Coleman M, Brown GC (2018) Neuronal cell death. *Physiological Reviews* 98: 813–880.

Frisch SM, Francis H (1994) Disruption of epithelial cell-matrix interactions induces apoptosis. *Journal of Cell Biology* 124: 619–626.

Fritz J, VoPham T, Wright KP, et al. (2020) A chronobiological evaluation of the acute effects of daylight saving time on traffic accident risk. *Current Biology* 30: 729–735.

Frost R (1914) Mending Wall. In: *North of Boston*. David Nutt, pp. 11–13.

Fuchs Y, Steller H (2011) Programmed cell death in animal development and disease. *Cell* 147: 742–758.

Fu Q, Posth C, Hajdinjak M, et al. (2016) The genetic history of Ice Age Europe. *Nature* 534: 200–205.

Furness JB, Bravo DM (2015) Humans as cucinivores: comparisons with other species. *Journal of Comparative Physiology B* 185: 825–834.

Gajardo-Vidal A, Lorca-Puls DL, PLORAS team, et al. (2021) Damage to Broca's area does not contribute to long-term speech production outcome after stroke. *Brain* 144: 817–832.

Gandhi AV, Mosser EA, Oikonomou G, et al. (2015) Melatonin is required for the circadian regulation of sleep. *Neuron* 85: 1193–1199.

Gandhi MK (1959) *The Moral Basis of Vegetarianism*. Navajiban Publishing House.

Garcia MA, Nelson WJ, Chavez N (2018) Cell-cell junctions organize structural and signaling networks. *Cold Spring Harbor Perspectives in Biology* 10(4): a029181.

Gelman R, Gallistel CR (2004) Language and the origin of numerical concepts. *Science* 306: 441–443.

Gervais M, Wilson DS (2005) The evolution and functions of laughter and humor: a synthetic approach. *Quarterly Review of Biology* 80: 395–430.

Ghavami S, M Hashemi M, Ande SR, et al. (2009) Apoptosis and cancer: mutations within caspase genes. *Journal of Medical Genetics* 46: 497–510.

Gibbons A (2014) The evolution of diet. *National Geographic* 226: 30–61.

Gibellini L, Moro L (2021) Programmed cell death in health and disease. *Cells* 10(7): 1765.

Gifford R (2014) Environmental psychology matters. *Annual Review of Psychology* 65: 541–579.

Gifford RJ (2021) Mapping the evolution of bornaviruses across geological timescales. *Proceedings of the National Academy of Sciences USA* 118(26): e2108123118.

Gilaberte Y, Prieto-Torres L, Pastushenko I, et al. (2016) Anatomy and function of the skin. In: *Nanoscience in Dermatology* (Hamblin MR, Avci P, Prow TW, eds). Academic Press, pp. 1–14.

Gilbert JA, Blaser MJ, Caporaso JG, et al. (2018) Current understanding of the human microbiome. *Nature Medicine* 24: 392–400.

Gillam E (2011) An introduction to animal communication. *Nature Education Knowledge* 3(10): 70.

Gilmore A (2005) Anoikis. *Cell Death Differentiation* 12: 1473–1477.

Giribet G, Sharma PP (2015) Evolutionary biology of harvestmen (Arachnida, Opiliones). *Annual Review of Entomology* 60: 157–175.Gould SJ (1990) *Wonderful Life: The Burgess Shale and the Nature of History.* WW Norton.

GlobalData (2022) United States of America (USA) skincare market size by categories, distribution channel, market share and forecast, 2021–2026. *GlobalData.* Retrieved from: https://www.globaldata.com/store/report/usa-skincare-market-analysis/. November 8, 2023.

Goodrum F (2016) Human cytomegalovirus latency: approaching the Gordian knot. *Annual Review of Virology* 3: 333–357.

Gould SJ (1990) *Wonderful Life: The Burgess Shale and the Nature of History.* W.W. Norton.

Gračanin A, Bylsma LM, Vingerhoets AJJM (2018) Why only humans shed emotional tears: evolutionary and cultural perspectives. *Human Nature* 29: 104–133.

Grandi N, Tramontano E (2018) Human endogenous retroviruses are ancient acquired elements still shaping innate immune responses. *Frontiers in Immunology* 9: 2039.

Graziano MSA (2022) The origin of smiling, laughing, and crying: the defensive mimic theory. *Evolutionary Human Sciences* 4: e10.

Grice EA, Segre JA (2011) The skin microbiome. *Nature Reviews Microbiology* 9: 244–253.

Griethuijsen LI, Trimmer BA (2014) Locomotion in caterpillars. *Biological Reviews Cambridge Philosophical Society* 89: 656–670.

Gross M (2021) Life underground. *Current Biology* 31: R415–R417.

Guerra KC, Crane JS (2023) *Sunburn. StatPearls [Internet].* StatPearls Publishing. Retrieved from: https://www.ncbi.nlm.nih.gov/books/NBK534837/. November 8, 2023.

Guilleminault C, Kirisoglu C, da Rosa AC, et al. (2006) Sleepwalking, a disorder of NREM sleep instability *Sleep Medicine* 7: 163–170.

Guy GP, Jr, Machlin SR, Ekwueme DU, Yabroff KR (2015) Prevalence and costs of skin cancer treatment in the U.S., 2002–2006 and 2007–2011. *American Journal of Preventive Medicine* 48: 183–187.

Hafen BB, Burns B (2021) *Physiology, Smooth Muscle.* National Center for Biotechnology Information. Retrieved from: https://www.ncbi.nlm.nih.gov/books/NBK526125/. November 1, 2023.

Hall MJR, Daniel M-V (2019) Visualization of insect metamorphosis. *Philosophical Transactions of the Royal Society B* 374: 20190071.

Han S, Lu Y, Xie J, et al. (2021) Probiotic gastrointestinal transit and colonization after oral administration: a long journey. *Frontiers in Cellular and Infection Microbiology* 11: 609722. https://www.frontiersin.org/articles/10.3389/fcimb.2021.609722/full. November 2, 2023.

Harada CN, Natelson Love MC, Triebel K (2013) Normal cognitive aging. *Clinical Geriatric Medicine* 29: 737–752.

Hardigan MA, Laimbeer FPE, Newton L, et al. (2017) Genome diversity of tuber-bearing Solanum uncovers complex evolutionary history and targets of domestication in the cultivated potato. *Proceedings of the National Academy of Sciences USA* 114: E9999–E10008.

Hardison RC (2012) Evolution of hemoglobin and its genes. *Cold Spring Harbor Perspectives in Medicine* 2(12): a011627.

Harrison SC (2008) Viral membrane fusion. *Nature Structural and Molecular Biology* 15: 690–698.

Harvey W (1628) De motu cordis (1628). In: The Circulation of the Blood and Other Writings (Franklin KJ, trans) [1957]. Richard Lowndes, p. 59.

Hawks J, Wang ET, Cocharan GM, et al. (2007) Recent acceleration of human adaptive evolution. *Proceedings of the National Academy of Sciences USA* 104: 20753–20758.

Hawksworth DL, Grube M (2020) Lichens redefined as complex ecosystems. *New Phytologist* 227: 1281–1283.

Hayashi T, Tokihiro T, Kurihara H, et al. (2017) Community effect of cardiomyocytes in beating rhythms is determined by stable cells. *Scientific Reports* 7: 15450.

Hayat H, Marmelshtein A, Krom AJ, et al. (2022) Reduced neural feedback signaling despite robust neuron and gamma auditory responses during human sleep. *Nature Neuroscience* 25: 935–943.

Hayflick L (1965) The limited *in vitro* lifetime of human diploid cell strains. *Experimental Cell Research* 37: 614–636.

Hazan R, Que YA, Maura D, et al. (2016) Auto poisoning of the respiratory chain by a quorum-sensing-regulated molecule favors biofilm formation and antibiotic tolerance. *Current Biology* 26: R80–R82.

Hebert PDN, Ratnasingham S, deWaard JR (2003) Barcoding animal life: cytochrome c oxidase subunit 1 divergences among closely related species. *Proceedings of the Royal Society of London B (Suppl.)* 270: S96–S99.

Hedrich R, Neher E (2018) Venus flytrap: how an excitable, carnivorous plant works. *Trends in Plant Science* 23: 220–234.

Heng C (2016) Tooth decay is the most prevalent disease. *Federal Practitioner* 33: 31–33.

Henry AG (2017) Neanderthal cooking and the costs of fire. *Current Anthropology* 58(Suppl 16): S329–S336.

Henry AG, Brooks AS, Piperno DR (2014) Plant foods and the dietary ecology of Neanderthals and early modern humans. *Journal of Human Evolution* 69: 44–54.Herben T, and Klimešová J (2020) Evolution of clonal growth forms in angiosperms. *New Phytologist* 225: 999–1010.

Herculano-Houzel S (2012) The remarkable, yet not extraordinary, human brain as a scaled-up primate brain and its associated cost. *Proceedings of the National Academy of Sciences USA* 109(suppl 1): 10661–10668.

Hernandez-Martinez R, Covarrubias L (2011) Interdigital cell death function and regulation: new insights on an old programmed cell death model. *Development, Growth and Differentiation.* 53: 245–258.

Hirano A, Shi G, Jones CR, et al. (2016) A cryptochrome 2 mutation yields advanced sleep phase in humans. *eLife* 5: e16695.

Hobaiter C, Byrne RW (2011) The gestural repertoire of the wild chimpanzee. *Animal Cognition* 14: 745–767.

Hobson JA (2005) Sleep is of the brain, by the brain and for the brain. *Nature* 437: 1254–1256.

Hodges RR, Dartt DA (2013) Tear film mucins: front line defenders of the ocular surface; comparison with airway and gastrointestinal tract mucins. *Experimental Eye Research* 117: 62–78.

Hohmann T, Dehghani F (2019) The cytoskeleton—a complex interacting meshwork. *Cells* 8: 362.

Holmberg KV, Hoffman MP (2015) Anatomy, biogenesis and regeneration of salivary glands. *Monographs in Oral Science* 24: 1–13.

Holowka NB, Lieberman DE (2018) Rethinking the evolution of the human foot: insights from experimental research. *Journal of Experimental Biology* 221: jeb174425.

Holstein TW, Laudet V (2014) Life-history evolution: at the origins of metamorphosis. *Current Biology* 24: R159–R161.

Honeybee Genome Sequencing Consortium (2006) Insights into social insects from the genome of the honeybee *Apis mellifera*. *Nature* 443: 931–949.

Honigmann A, Pralle A (2016) Compartmentalization of the cell membrane. *Journal of Molecular Biology* 428: 4739–4748.

Hopwood N (2022) "Not birth, marriage or death, but gastrulation": the life of a quotation in biology. *British Journal for the History of Science* 55: 1–26.

Horton CL (2017) Consciousness across sleep and wake: discontinuity and continuity of memory experiences as a reflection of consolidation processes. *Frontiers in Psychiatry* 8: 159. https://doi.org/10.3389/fpsyt.2017.00159.

Horvath S, Haghani A, Macoretta N, et al. (2022) DNA methylation clocks tick in naked mole rats but queens age more slowly than nonbreeders. *Nature Aging* 46: 46–59.

Hsia CC, Schmitz A, Lambertz M, et al. (2013) Evolution of air breathing: oxygen homeostasis and the transitions from water to land and sky. *Comparative Physiology* 3: 849–915.

Huber BA (2018) Cave-dwelling pholcid spiders (Araneae, Pholcidae): a review. *Subterranean Biology* 26: 1–18.

Hubisz MJ, Pollard KS (2014) Exploring the genesis and functions of human accelerated regions sheds light on their role in human evolution. *Current Opinion in Genetics and Development* 29: 15–21.

Hudson GV (1895) On seasonal time-adjustment in countries south of lat. 30°. *Transactions and Proceedings of the New Zealand Institute* 28: 734.

Huffard, CL (2006). Locomotion by *Abdopus aculeatus* (Cephalopoda: Octopodidae): walking the line between primary and secondary defenses. *Journal of Experimental Biology* 209: 3697–3707.

Huffard CL, Boneka F, Full RJ (2005) Underwater bipedal locomotion by octopuses in disguise. *Science* 307: 1927

Humphrey SP, Williamson RT (2001) A review of saliva: normal composition, flow, and function. *Journal of Prosthetic Dentistry* 85: 162–169.

Ierodiakonou D, Garcia-Larsen V, Logan A, et al. (2016) Timing of allergenic food introduction to the infant diet and risk of allergic or autoimmune disease: a systematic review and meta-analysis. *Journal of the American Medical Association* 316: 1181–1192.

Iosif A, Ballon B (2005) Bad moon rising: the persistent belief in lunar connections to madness. *Canadian Medical Association Journal* 173: 1498–1500.

Ishizuya-Oka A, Hasebe T, Shi YB (2010) Apoptosis in amphibian organs during metamorphosis. *Apoptosis* 15: 350–364.

Jabr F (2020) Why soap works. *New York Times*. Retrieved from: https://www.nytimes.com/2020/03/13/health/soap-coronavirus-handwashing-germs.html. November 1, 2023.

Jadhav U, Saxena M, O'Neill NK, et al. (2017) Dynamic reorganization of chromatin accessibility signatures during dedifferentiation of secretory precursors into Lgr5+ intestinal stem cells. *Cell Stem Cell* 21: 65–77.

James KR, Haritos N, Ades PK (2006) Mechanical stability of trees under dynamic loads. *American Journal of Botany* 93: 1522–1530.

Janzen HH (2004) Carbon cycling in earth systems—a soil science perspective. *Agriculture, Ecosystems and Environment* 104: 399–417.

Jezkova T, Wiens JJ (2017) What explains patterns of diversification and richness among animal phyla? *American Naturalist* 189: 201–212.

Johnson KG, Malow BA (2022) Daylight saving time: neurological and neuropsychological implications. *Current Sleep Medicine Reports* 8: 86–96.

Jorgensen I, Rayamajhi, Miao E (2017) Programmed cell death as a defence against infection. *Nature Reviews Immunology* 17: 151–164.

Jurisicova A, Acton BM (2004) Deadly decisions: the role of genes regulating programmed cell death in human preimplantation embryo development. *Reproduction* 128: 281–291.

Kalappurakkal JM, Sil P, Mayor S (2020) Toward a new picture of the living plasma membrane. *Protein Science* 29: 1355–1365.

Kamberov YG, Guhan SM, DeMarchis A, et al. (2008) Comparative evidence for the independent evolution of hair and sweat gland traits in primates. *Journal of Human Evolution* 125: 99–105.

Kannan MS, Lenca N (2012) *Field Guide to Algae and Other "Scums" in Ponds, Lakes, Streams and Rivers*. Booklet published by the Boone, Kenton and Campbell County Conservation Districts, KY, 5. Retrieved from: lp_kentucky-algae-guide.pdf (vermont.gov). May 16, 2024.

Kaplan S (1995) The restorative benefits of nature: towards an integrative framework. *Journal of Environmental Psychology* 15: 169–182.

Karlsson E, Kwiatkowski D, Sabeti P (2014) Natural selection and infectious disease in human populations. *Nature Reviews Genetics* 15: 379–393.

Katori M, Shi S, Ode KL, et al. (2022) The 103,200-arm acceleration dataset in the UK Biobank revealed a landscape of human sleep phenotypes. *Proceedings of the National Academy of Sciences USA* 119: e2116729119.

Kaur H, McDuff D, Williams AC, et al. (2022) "I didn't know I looked angry": characterizing observed emotion and reported affect at work. In *CHI Conference on Human Factors in Computing Systems (CHI '22)*, April 29–May 5, 2022, New Orleans, LA, USA. ACM, New York, NY. https://doi.org/10.1145/3491102.3517453.

Kawasaki J, Kojima S, Mukai Y, et al. (2021) 100-My history of bornavirus infections hidden in vertebrate genomes. *Proceedings of the National Academy of Sciences USA* 118(20): e2026235118.

Keene AD, Duboue ER (2018) The origins and evolution of sleep. *Journal of Experimental Biology* 221: jeb159533.

Kennedy KM, Bellissimo CJ, Breznik JA, et al. (2021) Over-celling fetal microbial exposure. *Cell* 184: P5839–5841.

Kerr JF, Wyllie AH, Currie AR (1972) Apoptosis: a basic biological phenomenon with wide-ranging implications in tissue kinetics. *British Journal of Cancer* 26: 239–527.

Ketelut-Carneiro N, Fitzgerald KA (2022) Apoptosis, pyroptosis, and necroptosis—oh my! The many ways a cell can die. *Journal of Molecular Biology* 434: 167378.

Khandia R, Dadar M, Munjal A, et al. (2019) A comprehensive review of autophagy and its various roles in infectious, non-infectious, and lifestyle diseases: current knowledge and prospects for disease prevention, novel drug design, and therapy. *Cells* 8: 674.

Kim T, Shin Y, Kang K, et al. (2022) Ultrathin crystalline-silicon-based strain gauges with deep learning algorithms for silent speech interfaces. *Nature Communications* 13: 5815.

Kistler L, Maezumi SY, Gregorio de Souza J, et al. (2018) Multiproxy evidence highlights a complex evolutionary legacy of maize in South America. *Science* 362: 1309–1313.

Knee K (nd) *Anthophyta*: evolution and diversity. Ohio State University. Retrieved from: https://plantfacts.osu.edu/resources/hcs300/angio1.htm. November 1, 2023.

Knöfler M, Haider S, Saleh L, et al. (2019) Human placenta and trophoblast development: key molecular mechanisms and model systems. *Cellular and Molecular Life Sciences* 76: 3479–3496.

Knowles TPJ, Vendruscolo M, Dobson CM (2014) The amyloid state and its association with protein misfolding diseases. *Nature Reviews Molecular Cell Biology* 15: 384–396.

Koch C (2018) What is consciousness? *Nature* 557: S8–S12.

Konopka RJ, Benzer S (1971) Clock mutants of *Drosophila melanogaster*. *Proceedings of the National Academy of Sciences USA* 68: 2112–2116.

Korb E, Finkbeiner S (2011) Arc in synaptic plasticity: from gene to behavior. *Trends in Neuroscience* 34: 591–598.

Koshiba-Takeuchi K, Mori AD, Kaynak BL, et al. (2009) Reptilian heart development and the molecular basis of cardiac chamber evolution. *Nature* 461: 95–98.

Krammer PH, Arnold R Lavrik IN (2007) Life and death in peripheral T cells. *Nature Reviews Immunology* 7: 532–542.

Krieger R (2010) *Dissecting a chicken leg. Krieger Science Blog.* Retrieved from: https://www.krie egerscience.com/anatomy/chicken-leg/. November 1, 2023.

Krishnana Y, Grodzinsky AJ (2018) Cartilage diseases. *Matrix Biology* 71–72: 51–69.

Kroemer G, Galluzzi L, Vandenabeele P, et al. (2016) Programmed cell death 50 (and beyond). *Cell Death and Differentiation* 23: 10–17.

Krumhuber EG, Manstead ASR (2009) Can Duchenne smiles be feigned? New evidence on felt and false smiles. *Emotion* 9: 807–820.

Kucik CJ, Martin GL, Sortor BV (2004) Common intestinal parasites. *American Family Physician* 69: 1161–1168.

Kuhl P (2004) Early language acquisition: cracking the speech code. *Nature Reviews Neuroscience* 5: 831–843.

Kun E, Javan EM, Smith O, et al. (2023) The genetic architecture and evolution of the human skeletal form. *Science* 381: eadf8009.

Kuo, M (2015) How might contact with nature promote human health? Promising mechanisms and a possible central pathway. *Frontiers in Psychology* 6: 1093. https://doi.org/10.3389/fpsyg.2015.01093.

Lacey N, Kavanagh K, Tseng SC (2009) Under the lash: Demodex mites in human diseases. *Biochemistry (London)* 31: 2–6.

Lacey N, Raghallaigh SN, Powell FC (2011) Demodex mites—commensals, parasites or mutualistic organisms? *Dermatology* 222: 128–130.

Lameira AR, Moran S (2023) Life of p: a consonant older than speech. *BioEssays* 22023: 200246. https://doi.org/10.1002/bies.202200246.

Lang F, Gulbins E, Lang PA, Zappulla D, Föller M (2010) Ceramide in suicidal death of erythrocytes. *Cellular Physiology and Biochemistry* 26: 21–28.

Langley RL, Morrow WE (1997) Deaths resulting from animal attacks in the United States. *Wilderness and Environmental Medicine* 8: 8–16.

Larragoite ET, Spivak AM (2019) Viral latency: down but not out. *eLife* 8: e53363.

Latinus M, Belin P (2011) Human voice perception. *Current Biology* 21: R143–R145.

Laumer CE, Fernández R, Lemer S, et al. (2019) Revisiting metazoan phylogeny with genomic sampling of all phyla. *Proceedings of the Royal Society B* 286: 20190831.

Lawrie RA (2006) The conversion of muscle to meat. In: *Lawrie's Meat Science*, 7th edition. Woodhead Publishing, pp. 128–156.

Lederberg J, McCray AT (2001) "Ome sweet omics"—a genealogical treasury of words. *Scientist* 15: 8–9.

Lee A (2001) Membrane structure. *Current Biology* 11: R811–R814.

Leggett HC, Cornwallis CK, West SA (2012) Mechanisms of pathogenesis, infective dose and virulence in human parasites. *Public Library of Science Pathogens* 8(2): e1002512.

Lepczyk CA, La Sorte FA, Aronson MFJ, et al. (2017) Global patterns and drivers of urban bird diversity. In: *Ecology and Conservation of Birds in Urban Environments* (Murgui E and Hedblom M, eds). Springer, pp. 13–33.

Leung AKC, Leung AAM, Wong AHC, et al. (2020) Sleep terrors: an updated review. *Current Pediatric Review* 16: 176–182.

Lidsky PV, Andino R (2020) Epidemics as an adaptive driving force determining lifespan setpoints. *Proceedings of the National Academy of Sciences USA* 117: 17937–17948.

Lieberman P (2007) The evolution of human speech: its anatomical and neural bases. *Current Anthropology* 48: 39–66.

Lieberson AD (2004) How long can a person survive without food? *Scientific American.* Retrieved from: https://www.scientificamerican.com/article/how-long-can-a-person-survive-without-food/. November 2, 2023.

Liggan LM, Martone PT (2018) Under pressure: biomechanical limitations of developing pneumatosysts in the bull kelp (*Nereocystis luetkeana*, Phaeophyceae). *Journal of Phycology* 54: 608–615.

Li H, Tian Y, Menolli N, et al. (2021) Reviewing the world's edible mushroom species: a new evidence-based classification system. *Comprehensive Reviews in Food Science and Food Safety* 20: 1982–2014.

Lindsey R, Dahlman L (2023) *Climate change: Global temperature.* Climate.gov. NOAA. Retrieved from: https://www.climate.gov/news-features/understanding-climate/climate-change-global-temperature. August 13, 2023.

Lingle S, Wyman MT, Kotraba R, et al. (2012) What makes a cry a cry? A review of infant distress vocalizations. *Current Zoology* 58: 698–726.

Liscum E, Askinosie SK, Leuchtman DL, et al. (2014) Phototropism: growing towards an understanding of plant movement. *Plant Cell* 26: 38–55.

Liu B, Tai Y, Achanta S, et al. (2016) IL-33/ST2 signaling excites sensory neurons and mediates itch response in a mouse model of poison ivy contact allergy. *Proceedings of the National Academy of Sciences USA* 113: E7572–E7579.

Liu C, Gao J, Cui X, et al. (2021) A towering genome: experimentally validated adaptations to high blood pressure and extreme stature in the giraffe. *Science Advances* 7: eabe9459.

Liu J, Mosti F, Silver DL (2021) Human brain evolution: emerging roles for regulatory DNA and RNA. *Current Opinion in Neurobiology* 71: 170–177.

Lockshin RA (2016) Programmed cell death 50 (and beyond). *Cell Death Differentiation* 23: 10–17.

Lockshin RA, Williams CM (1965) Programmed cell death: V. Cytolytic enzymes in relation to the breakdown of the intersegmental muscles of silkmoths. *Journal of Insect Physiology* 11: 831–844.

Lockshin R, Zakeri Z (2004) Caspase-independent cell death? *Oncogene* 23: 2766–2773.

Loftfield E, Freedman ND, Dodd KW, et al. (2016) Coffee drinking is widespread in the United States, but usual intake varies by key demographic and lifestyle factors. *Journal of Nutrition* 146: 1762–1768.

Lopez MJ, Mohiuddin SS (2021) Biochemistry, essential amino acids. *StatPearls.* Retrieved from: https://www.ncbi.nlm.nih.gov/books/NBK557845/. November 2, 2023.

Lummaa V, Vuorisalo T, Barr RG, et al. (1998) Why cry? Adaptive significance of intensive crying in human infants. *Evolution and Human Behavior* 19: 93–202.

Lutzoni LF, Miadlikowska J (2009) Lichens. *Current Biology* 19: R502–R503.

Lu W, Meng QJ, Tyler N, et al. (2010) A circadian clock is not required in an Arctic mammal. *Current Biology* 20: 533–537.

MacPherson K (2010) *The "sultan of slime": biologist continues to be fascinated by organisms after nearly 70 years of study.* Princeton University. Retrieved from: https://www.princeton.edu/news/2010/01/21/sultan-slime-biologist-continues-be-fascinated-organisms-after-nearly-70-years. November 2, 2023.

Maderson PFA (2003) Mammalian skin evolution: a reevaluation. *Experimental Dermatology* 12: 233–236.

Madison-Antenucci S, Kramer LD, Gebhardt LL, et al. (2020) Emerging tick-borne diseases. *Clinical Microbiology Review* 33(2): e00083-18.

Maghsoodi A, Chatterjee A, Andricioaei I, et al. (2019) How the phage T4 injection machinery works including energetics, forces, and dynamic pathway. *Proceedings of the National Academy of Sciences USA* 116: 25097–25105.

Magnabosco C, Lin L-H, Dong H, et al. (2018) The biomass and biodiversity of the continental subsurface. *Nature Geoscience* 11: 707–717.

Mahasen LMA (2016) Evolution of the kidney. *Anatomy Physiology Biochemistry International Journal* 1: 1–6.

Mahowald MW, Schenck CH (2005) Insights from studying human sleep disorders. *Nature* 437: 1279–1285.

Maldonado M, López-Acosta M, Sitjà C, et al. (2013) A giant foraminifer that converges to the feeding strategy of carnivorous sponges: *Spiculosiphon oceana* sp. nov. (Foraminifera, Astrorhizida). *Zootaxa* 3669: 571–584.

Maleszka R (2018) Beyond Royalactin and a master inducer explanation of phenotypic plasticity in honey bees. *Communications Biology* 1: 8. pmid:30271895.

Malorni W, Nuñez G, Peter ME, et al. (2009) Classification of cell death. *Cell Death and Differentiation* 16: 3–11.

Manfredini R, Fabbian F, Cappadona R, et al. (2019) Daylight saving time and acute myocardial infarction: a meta-analysis. *Journal of Clinical Medicine* 8: 404.

Mariz JP, Nery MF (2020) Unraveling the molecular evolution of blood coagulation genes in fishes and cetaceans. *Frontiers in Marine Science* 7: 592383. https://doi.org/10.3389/fmars.2020.592383.

Marques E, Chen TM (2023) Actinic keratosis. *StatPearls [Internet]*. StatPearls Publishing. Retrieved from: https://www.ncbi.nlm.nih.gov/books/NBK557401/. November 8, 2023.

Márquez-Ruiz J, Escudero M (2008) Tonic and phasic phenomena underlying eye movements during sleep in the cat. *Journal of Physiology* 586: 3461–3477.

Marshall M (2009) Timeline: the evolution of life. *NewScientist*. Retrieved from: https://www.newscientist.com/article/dn17453-timeline-the-evolution-of-life/. November 1, 2023.

Marshall M (nd) Natural selection. *New Scientist*. Retrieved from: https://www.newscientist.com/definition/natural-selection/. November 1, 2023.

Marsh PD (2006) Dental plaque as a biofilm and a microbial community—implications for health and disease. *BioMed Central Oral Health* 6: S14.

Matzinger T, Fitch WT (2021) Voice modulatory cues to structure across languages and species. *Philosophical Transactions of the Royal Society B* 376: 20200393.

Ma X, Cong P, Hou X, et al. (2014) An exceptionally preserved arthropod cardiovascular system from the early Cambrian. *Nature Communications* 5: 3560.

Mazzarello P (2000) What dreams may come? *Nature* 408: 523.

McCully ME (1999) Roots in soil: unearthing the complexities of roots and their rhizospheres. *Annual Review of Plant Physiology and Plant Molecular Biology* 50: 695–718.

McFall-Ngaia M, Hadfield MG, Bosch TCG, et al. (2013) Animals in a bacterial world, a new imperative for the life sciences. *Proceedings of the National Academy of Sciences USA* 110: 3229–3236.

McHill AW, Chinoy ED (2020) Utilizing the National Basketball Association's COVID-19 restart "bubble" to uncover the impact of travel and circadian disruption on athletic performance. *Science Reports* 10: 21827.

McNear, DH, Jr (2013) The rhizosphere—roots, soil and everything in between. *Nature Education Knowledge* 4(3): 1.

Medina-Sauza RM, Álvarez-Jiménez M, Delha A, et al. (2019) Earthworms building up soil microbiota, a review. *Frontiers in Environmental Science* 7: 81.

Medzhitov R, Janeway CA Jr (2002) Decoding the patterns of self and nonself by the innate immune system. *Science* 296: 298–300.

Mendelson CR, Gao L, Montalbano AP (2019) Multifactorial regulation of myometrial contractility during pregnancy and parturition. *Frontiers in Endocrinology* 10: 714. https://doi.org/10.3389/fendo.2019.00714.

Menon R, Bonney EA, Condon J, et al. (2016) Novel concepts on pregnancy clocks and alarms: redundancy and synergy in human parturition. *Human Reproduction Update* 22: 535–560.

Merhaut DJ (1999) How do large trees, such as redwoods, get water from their roots to the leaves? *Scientific American*. Retrieved from: https://www.scientificamerican.com/article/how-do-large-trees-such-a/. November 2, 2023.

Meyer RS, Purugganan MD (2013) Evolution of crop species: genetics of domestication and diversification. *Nature Reviews Genetics* 14: 840–852.

Miller SL, Urey HC (1959) Organic compound synthesis on the primitive earth. *Science* 130: 245–251.

Mishra A, Lai GC, Yao LJ, et al. (2021) Microbial exposure during early human development primes fetal immune cells. *Cell* 184: 3394–3409.

Mitchell R, Popham F (2008) Effect of exposure to natural environment on health inequalities: an observational population study. *Lancet* 372: 1655–1660.

Mohawk JA, Green CB, Takahashi JS (2012) Central and peripheral circadian clocks in mammals. *Annual Review of Neuroscience* 35: 445–462.

Monahan-Earley R, Dvorak AM, Aird WC (2013) Evolutionary origins of the blood vascular system and endothelium. *Journal of Thrombosis and Haemostasis* 11: 46–66.

Mondal S, Pramanik K, Panda D, et al. (2022) Sulfur in seeds: an overview. *Plants (Basel)* 11(3): 450.

Moore PS, Chang Y (2010) Why do viruses cause cancer? Highlights of the first century of human tumour virology. *Nature Reviews Cancer* 10: 878–889.

Morand S, Bouamer S, Hugot JP (2006) Nematodes. In: *Micromammals and Macroparasites: From Evolutionary Ecology to Management.* Tokyo: Springer Japan. pp. 63–79.

Moras M, Lefevre SD, Ostuni MA (2017) From erythroblasts to mature red blood cells: organelle clearance in mammals. *Frontiers in Physiology* 8: 1076.

Mordecai E, Cohen J, Evans MV, et al. (2017) Detecting the impact of temperature on transmission of Zika, dengue, and chikungunya using mechanistic models. *Public Library of Science Neglected Tropical Diseases* 11: e0005568.

Moss EG (2007) Heterochronic genes and the nature of developmental time. *Current Biology* 17: R425–R434.

Morrison SF, Nakamura K (2019) Central mechanisms for thermoregulation. *Annual Review of Physiology* 81: 285–308.

Moujalled D, Strasser A, Lidell JR (2021) Molecular mechanisms of cell death in neurological diseases. *Cell Death and Differentiation* 28: 2029–2044.

Moulia B, Douady S, Hamant O (2021) Fluctuations shape plants through proprioception. *Science* 372: eabc6868.

Muñoz-Fontela C, Dowling WE, Funnell SGP, et al. (2020) Animal models for COVID-19. *Nature* 586: 509–515.

Mustelin T, Ukadike KC (2020) How retroviruses and retrotransposons in our genome may contribute to autoimmunity in rheumatological conditions. *Frontiers in Immunology* 11: 593891. https://doi.org/10.3389/fimmu.2020.593891.

Mylona P, Pawlowski K, Bisseling T (1995) Symbiotic nitrogen fixation. *Plant Cell* 7: 869–885.

Nagahata Y, Masuda K, Nishimura Y, et al. (2022) Tracing the evolutionary history of blood cells to the unicellular ancestor of animals. *Blood* 140: 2611–2625.

NCHS (2015) *National Center for Health Statistics.* Retrieved from: https://www.cdc.gov/nchs/fastats/immunize.htm. November 1, 2023.

Neale DB, Wheeler NC (2019) The conifers. In: *The Conifers: Genomes, Variation and Evolution.* Springer. pp. 1–21.

Needelman BA (2013) What are soils? *Nature Education Knowledge* 4(3): 2.

Nekola JC (2014) Overview of the North American terrestrial gastropod fauna. *American Malacological Bulletin* 32: 225–235.

NewScientist (2009) The enemy within: 10 human parasites. *NewScientist.* Retrieved from: https://www.newscientist.com/gallery/mg20327161300-enemy-within-human-parasites/. November 2, 2023.

Nguyen TTM, Gillet G, Popgeorgiev N (2021) Caspases in the developing central nervous system: apoptosis and beyond. *Frontiers in Cell and Developmental Biology* 9: 702404. https://doi.org/10.3389/fcell.2021.702404.

NHLBI (2022) *What are sleep deprivation and deficiency?* National Heart, Lung, and Blood Institute. Retrieved from: https://www.nhlbi.nih.gov/health/sleep-deprivation#:~:text=

Sleep%20deficiency%20is%20linked%20to,adults%2C%20teens%2C%20and%20child ren. November 3, 2023.

NIH (2022) Melatonin: what you need to know. National Center for Complementary and Integrative Health. *National Institutes of Health*. Retrieved from: https://www.nccih.nih. gov/health/melatonin-what-you-need-to-know. November 2, 2023.

Niklas KJ, Kurschera U (2010) The evolution of the land plant life cycle. *The New Phytologist.* 185: 27–41.

Nishimura T, Tokuda IT, Miyachi S, et al. (2022) Evolutionary loss of complexity in human vocal anatomy as an adaptation for speech. *Science* 377: 760–763.

Ni W, Yang X, Yang D, et al. (2020) Role of angiotensin-converting enzyme 2 (ACE2) in COVID-19. *Critical Care* 24: 422.

Nogimura D, Mizushige T, Taga Y, et al. (2020) Prolyl-hydroxyproline, a collagen-derived dipeptide, enhances hippocampal cell proliferation, which leads to antidepressant-like effects in mice. *FASEB Journal* 34: 5715–5723.

Noto RE, Leavitt L, Edens MA (2021) Physiology, muscle. *StatPearls*. StatPearls Publishing. Retrieved from: https://www.ncbi.nlm.nih.gov/books/NBK532258/. November 1, 2023.

Nunn CL, Samson DR (2018) Sleep in a comparative context: investigating how human sleep differs from sleep in other primates. *American Journal of Biological Anthropology* 166: 601–612.

Ochs M, Nyengaard JR, Jung A, et al. (2004) The number of alveoli in the human lung. *American Journal of Respiratory Critical Care Medicine* 169: 120–124.

Oetjen LK, Mack MR, Feng J, et al. (2017) Sensory neurons co-opt classical immune signaling pathways to mediate chronic itch. *Cell* 171: 217–228.

Ong CC, Gopinath SCB, Rebecca LWX, et al. (2018) Diagnosing human blood clotting deficiency. *International Journal of Biological Macromolecules* 116: 765–773.

Opferman J (2008) Apoptosis in the development of the immune system. *Cell Death and Differentiation* 15: 234–242.

Opferman JT, Korsmeyer SJ (2003) Apoptosis in the development and maintenance of the immune system. *Nature Immunology* 4: 410–415.

Ouyang L, Shi Z, Zhao S, et al. (2012) Programmed cell death pathways in cancer: a review of apoptosis, autophagy and programmed necrosis. *Cell Proliferation* 45: 487–498.

Parham RA, Gray RL (1984) Formation and structure of wood. In: *Chemistry of Solid Wood* (Rowell R, ed), Adv. Chem. Series 207. American Chemical Society, pp. 3–56.

Park CH (2020) Cost-effective management of severe gastroesophageal reflux disease: toward an improved understanding of anti-reflux surgery. *Journal of Neurogastroenterology and Motility* 26: 169–170.

Parrish NF, Tomonaga K (2016) Endogenized viral sequences in mammals. *Current Opinion in Microbiology* 31: 176–183.

Parrish NF, Tomonaga K (2018) A viral (Arc)hive for metazoan memory. *Cell* 172: 8–10.

Pastuzyn ED, Day CE, Kearns RB, et al. (2018) The neuronal gene *Arc* encodes a repurposed retrotransposon Gag protein that mediates intercellular RNA transfer. *Cell* 172: 275–288.

Patino-Ramirez F, Arson C, Dussutour A (2021) Substrate and cell fusion influence on slime mold network dynamics. *Scientific Reports* 11: 1498.

Paul B, Sterner ZR, Buchholz DR, et al. (2022) Thyroid and corticosteroid signaling in amphibian metamorphosis. *Cells* 11: 1595. https://doi.org/10.3390/cells11101595.

Pedersen AML, Sørensen CE, Proctor GB, et al. (2018) Salivary secretion in health and disease. *Journal of Oral Rehabilitation* 45: 730–746.

Peever J, Fuller PM (2017) The biology of REM sleep. *Current Biology* 27: R1237–R1248.

Peñalver E, Arillo A, Delclòs X, et al. (2017) Ticks parasitised feathered dinosaurs as revealed by Cretaceous amber assemblages. *Nature Communications* 8: 1924.

Perreault C, Mathew S (2012) Dating the origin of language using phonemic diversity. *PLoS One* 7(4): e35289.

Perry G, Dominy N, Claw K, et al. (2007) Diet and the evolution of human amylase gene copy number variation. *Nature Genetic* 39: 1256–1260.

Persat A, Nadell CD, Kim MK, et al. (2015) The mechanical world of bacteria. *Cell* 161: 988–997.

Petersen MA, Ryu JK, Akassoglou K (2018) Fibrinogen in neurological diseases: mechanisms, imaging and therapeutics. *Nature Reviews Neuroscience* 19: 283–301.

Pickard GE, Sollars PJ (2012) Intrinsically photosensitive retinal ganglion cells. *Reviews in Physiology, Biochemistry and Pharmacology* 162: 59–90.

Pierce EC, Morin M, Little JC, et al. (2021) High-throughput genetic screen reveals diverse impacts of cheese-associated fungi on bacteria. *Nature Microbiology* 6: 87–102.

Pika S, Wilkinson R, Kendrick KH, et al. (2018) Taking turns: bridging the gap between human and animal communication. *Proceedings of the Royal Society B* 285: 20180598.

Pillai AS, Chandler SA, Liu Y, et al. (2020) Origin of complexity in haemoglobin evolution. *Nature* 581: 480–485.

Piperno DR, Ranere AJ, Holst I, et al. (2009) Starch grain and phytolith evidence for early ninth millennium B.P. maize from the Central Balsas River Valley, Mexico. *Proceedings of the National Academy of Sciences USA* 106: 5019–5024.

Pittendrigh CS (1993) Reflections of a Darwinian clock-watcher. *Annual Review of Physiology* 55: 17–54.

Plümper O, King HE, Geisler T, et al. (2017) Subduction zone forearc serpentinites as incubators for deep microbial life. *Proceedings of the National Academy of Sciences USA* 114: 4324–4329.

Pollard KS, Salama SR, King B, et al. (2006) Forces shaping the fastest evolving regions in the human genome. *Public Library of Science Genetics* 2(10): e168.

Pollard KS, Salama SR, Lambert N, et al. (2006) An RNA gene expressed during cortical development evolved rapidly in humans. *Nature* 443: 167–172.

Poppinga S, Böse A-S, Seidel R, et al. (2019) A seed flying like a bullet: ballistic seed dispersal in Chinese witch-hazel (*Hamamelis mollis* OLIV., Hamamelidaceae). *Journal of the Royal Society Interface* 16: 20190327.

Poroyko V, Carreras A, Khalyfa A, et al. (2016) Chronic sleep disruption alters gut microbiota, induces systemic and adipose tissue inflammation and insulin resistance in mice. *Scientific Reports* 6: 35405.

Porter SM (2007) Seawater chemistry and early carbonate biomineralization. *Science* 316: 1302.

Prather AA, Janicki-Deverts D, Hall MH, et al. (2015) Behaviorally assessed sleep and susceptibility to the common cold. *Sleep* 38: 1353–1359.

Prats-Uribe A, Tobías A, Prieto-Alhambra D (2018) Excess risk of fatal road traffic accidents on the day of daylight saving time change. *Epidemiology* 29: e44–e45.

Pretorius E, du Plooy JN, Bester J (2016) A comprehensive review on eryptosis. *Cellular Physiology and Biochemistry* 39: 1977–2000.

Proix T, Delgado Saa J, Christen A, et al. (2022) Imagined speech can be decoded from low-and cross-frequency intracranial EEG features. *Nature Communications* 13: 1–14.

Pruszynski JA, Johansson RS (2014). Edge-orientation processing in first-order tactile neurons. *Nature Neuroscience* 17: 1404–1409.

Qui Y-L, Li L, Wang B, et al. (2006) The deepest divergences in land plants inferred from phylogenomic evidence. *Proceedings of the National Academy of Sciences USA* 103: 15511–15516.

Rackaityte E, Halkias J, Fukui EM, et al. (2020) Viable bacterial colonization is highly limited in the human intestine in utero. *Nature Medicine* 26: 599–607.

Rajjou L, Duval MM, Gallardo-Guerrero KK, et al. (2012) Seed germination and vigor. *Annual Review of Plant Biology* 63: 507–533.

Raposo AC, Portela RD, Aldrovani M, et al. (2020) Comparative analysis of tear composition in humans, domestic mammals, reptiles, and birds. *Frontiers in Veterinary Science* 7: 283.

Rauma M, Boman A, Johanson G (2013) Predicting the absorption of chemical vapours. *Advances in Drug Deliver Reviews* 65: 306–314.

Ravindran S (2012) Barbara McClintock and the discovery of jumping genes. *Proceedings of the National Academy of Sciences USA* 109: 20198–20199.

Razani N, Morshed S, Kohn MA, et al. (2018) Effect of park prescriptions with and without group visits to parks on stress reduction in low-income parents: SHINE randomized trial. *Public Library of Science One* 13(2): e0192921.

Reddy S, Reddy V, Sharma S (2023) Physiology, circadian rhythm. [Updated May 1, 2023]. In: StatPearls. StatPearls Publishing. Retrieved from: https://www.ncbi.nlm.nih.gov/books/NBK519507/#:~:text=Circadian%20rhythm%20is%20the%2024;Earth's%20rotation%20around%20its%20axis. November 2, 2023.

Reinhart BJ, Slack FJ, Basson M, et al. (2000) The 21-nucleotide let-7 RNA regulates developmental timing in *Caenorhabditis elegans*. *Nature* 403: 901–906.

Rensing SA (2018) Great moments in evolution: the conquest of land by plants. *Annual Review of Plant Biology* 42: 49–54.

Reppert SM, de Roode JC (2018) Demystifying monarch butterfly migration. *Current Biology*: 28: R1009–R1022.

Reuben C, Elgaddal N, Black LI (2023) *Sleep medication use in adults aged 18 and over: United States, 2020. National Center for Health Statistics Data Brief No. 462.* Retrieved from: https://www.cdc.gov/nchs/products/databriefs/db462.htm. November 3, 2023.

Richard D, Liu Z, Cao J, et al. (2020) Evolutionary selection and constraint on human knee chondrocyte regulation impacts osteoarthritis risk. *Cell* 181: 362–381.

Richards MP, Trinkaus E (2009) Isotopic evidence for the diets of European Neanderthals and early modern humans. *Proceedings of the National Academy of Sciences USA* 106: 16034–16039.

Richardson K, Steffen W, Lucht W, et al. (2023) Earth beyond six of nine planetary boundaries. *Science Advances* 9: eadh2458.

Rich MK, Vigneron N, Libourel C, et al. (2021) Lipid exchanges drove the evolution of mutualism during plant terrestrialization. *Science* 372: 864–868.

Roberts NF (2022) Despite $65 billion a year sleep aid market, Americans remain sleep deprived. *Forbes.* Retrieved from: https://www.forbes.com/sites/nicoleroberts/2022/03/20/despite-65-billion-a-year-sleep-aid-market-americans-remain-sleep-deprived/?sh=3784e3b67521. November 3, 2023.

Robert VA, Casadevall A (2009) Vertebrate endothermy restricts most fungi as potential pathogens. *Journal of Infectious Diseases* 200: 1623–1626.

Roblegg E, Coughran A, Sirjani D (2019) Saliva: an all-rounder of our body. *European Journal of Pharmaceutics and Biopharmaceutics* 142: 133–141.

Rodari A, Darcis G, Van Lint CM (2021) The current status of latency reversing agents for HIV-1 remission. *Annual Review of Virology* 8: 491–514.

Rolff J, Johnston PR, Reynolds S (2019) Complete metamorphosis of insects. *Philosophical Transactions of the Royal Society of London* 374(1783): 20190063.

Romero A, De Juan J (2012) SEM, teeth and palaeoanthropology: the secret of ancient human diets. In: *Scanning Electron Microscopy for the Life Sciences* (Schatten H, ed). Cambridge University Press, pp. 236–256.

Romero A, Ramírez-Rozzi FV, De Juan J, et al. (2013) Diet-related buccal dental microwear patterns in Central African Pygmy foragers and Bantu-speaking farmer and pastoralist populations. *Public Library of Science One* 8: 2.

Rosenbaum DM, Rasmussen SG, Kobilka BK (2009) The structure and function of G-protein-coupled receptors. *Nature* 459: 356–363.

Rosental B, Kowarsky M, Seita J, et al. (2018) Complex mammalian-like haematopoietic system found in a colonial chordate. *Nature* 564: 425–429.

Ross MD, Owren MJ, Zimmermann E (2009) Reconstructing the evolution of laughter in great apes and humans. *Current Biology* 19: 1106–1111.

Rummer JL, McKenzie DJ, Innocenti A, et al. (2013) Root effect hemoglobin may have evolved to enhance general tissue oxygen delivery. *Science* 340: 1327–1329.

Sabato G (2019) What's so funny? The science of why we laugh. *Scientific American*. Retrieved from: https://www.scientificamerican.com/article/whats-so-funny-the-science-of-why-we-laugh/. November 2, 2023.

Sacks H, Schegloff EA, Jefferson G (1974) A simplest systematics for the organization of turn-taking in conversation. *Language* 50: 696–735.

Sagan L (1967) On the origin of mitosing cells. *Journal of Theoretical Biology* 14: 225–274.

Saheb Kashaf S, Proctor DM, Deming C, et al. (2022) Integrating cultivation and metagenomics for a multi-kingdom view of skin microbiome diversity and functions. *Nature Microbiology* 7: 169–179.

Saini R, Saini S, Sharma S (2011) Biofilm: a dental microbial infection. *Journal of Natural Science Biology and Medicine* 2: 71–75.

Sakamoto K, Hondo Y, Takahashi N, et al. (2021) Emergent synchronous beating behavior in spontaneous beating cardiomyocyte clusters. *Scientific Reports* 11: 11869.

Salehi S, Scheibel T (2018) Biomimetic spider silk fibres: from vision to reality. *Biochemistry* 40: 4–7.

Saper CB, Scammell TE, Lu J (2005) Hypothalamic regulation of sleep and circadian rhythms. *Nature* 437: 1257–1263.

Savin T, Kurpios NA, Shyer AE, et al. (2011) On the growth and form of the gut. *Nature* 476: 57–62.

Scammell TE (2015) Narcolepsy. *New England Journal of Medicine* 373: 2654–2662.

Scannapieco FA, Cantos A (2016) Oral inflammation and infection, and chronic medical diseases: implications for the elderly. *Periodontology 2000* 72: 153–175.

Schatz O, Langer E, Ben-Arie N (2014) Gene dosage of the transcription factor Fingerin (bHLHA9) affects digit development and links syndactyly to ectrodactyly. *Human Molecular Genetics* 23: 5394–5401.

Schmidt C (2008) Phylogeny of the terrestrial Isopoda (Oniscidea): a review. *Arthropod Systematics and Phylogeny* 66: 191–226.

Schmuth M, Martinz V, Janecke A, et al. (2013) Inherited ichthyoses/generalized Mendelian disorders of cornification. *European Journal of Human Genetics* 21: 123–133.

Schoeneberger PJ, Wysocki DA, Benham EC, Soil Survey Staff (2012) *Field Book for Describing and Sampling Soils, Version 3.0*. Natural Resources Conservation Service, National Soil Survey Center.

Schulte K, Kunter U, Moeller MJ (2015) The evolution of blood pressure and the rise of mankind. *Nephrology Dialysis Transplantation* 30: 713–723.

Schwander M, Kachar B, Müller U (2010) The cell biology of hearing. *Journal of Cell Biology* 190: 9–20.

Schwartz RM, Dayhoff MO (1978) Origins of prokaryotes, eukaryotes, mitochondria, and chloroplasts: a perspective is derived from protein and nucleic acid sequence data. *Science* 199: 395–403.

Scitable (2014) Protein structure. *Nature Education*. Retrieved from: https://www.nature.com/scitable/topicpage/protein-structure-14122136/. November 1, 2023.

Scudellari M (2021) How the coronavirus infects cells—and why Delta is so dangerous. *Nature* 595: 640–644.

Seafood Source (nd) *Seafood Handbook*. Retrieved from: https://www.seafoodsource.com/seafood-handbook. November 1, 2023.

Searcy WA, Nowicki S (2005) *The Evolution of Animal Communication*. Princeton University Press.

Sehgal A (2017) Physiology flies with time. *Cell* 171: 1232–1235.

Sekizawa A, Samura O, Zhen DK, et al. (2000) Apoptosis in fetal nucleated erythrocytes circulating in maternal blood. *Prenatal Diagnosis* 20: 886–889.

Sen Gupta A (2019) Hemoglobin-based oxygen carriers: Current state-of-the-art and novel molecules. *Shock* 52(1S Suppl 1): 70–83.

Sender R, Fuchs S, Milo R (2016) Revised estimates for the number of human and bacteria cells in the body. *Public Library of Science Biology* 14(8): e1002533.

Setlow P (2003) Spore germination. *Current Opinion in Microbiology* 6: 550–556.

Seymour RS (2016) Cardiovascular physiology of dinosaurs. *Physiology* 31: 430–441.

Shain A, Bastian B (2016) From melanocytes to melanomas. *Nature Reviews Cancer* 16: 345–358.

Shalini S, Dorstyn L, Dawar S, et al. (2015) Old, new and emerging functions of caspases. *Cell Death Differentiation* 22: 526–539.

Sharifi R, Ryu C-M (2021) Social networking in crop plants: wired and wireless cross-plant communications. *Plant Cell Environment* 44: 1095–1110.

Sharma A, Boise LH, Shanmugam M (2019) Cancer metabolism and the evasion of apoptotic cell death. *Cancers* 11: 1144.

Sharma AN, Patel BC (2023) *Laser Fitzpatrick skin type recommendations. [Updated March 6, 2023]. StatPearls [Internet].* StatPearls Publishing. Retrieved from: https://www.ncbi.nlm.nih.gov/books/NBK557626/. November 8, 2023.

Shepherd JD (2018) Arc—an endogenous neuronal retrovirus? *Seminars in Cell and Developmental Biology* 77: 73–78.

Shibata Y, Tanizaki Y, Zhang H, et al. (2021) Thyroid hormone receptor is essential for larval epithelial apoptosis and adult epithelial stem cell development but not adult intestinal morphogenesis during *Xenopus tropicalis* metamorphosis. *Cells* 10(3): 536.

Shibutani ST, Saitoh, T, Nowag H, et al. (2015). Autophagy and autophagy-related proteins in the immune system. *Nature Immunology* 16: 1014–1024.

Shogbesan O, Poudel DR, Victor S, et al. (2018) A systematic review of the efficacy and safety of fecal microbiota transplant for *Clostridium difficile* infection in immunocompromised patients. *Canadian Journal of Gastroenterology and Hepatology* 2018: 1394379.

Shu K, Liu X-d, Xie Q, et al. (2016) Two faces of one seed: hormonal regulation of dormancy and germination. *Molecular Plant* 9: 34–35.

Siegel JM (2008) Do all animals sleep? *Trends in Neurosciences* 31: 208–213.

Sigwart JD, Sutton MD (2007) Deep molluscan phylogeny: synthesis of palaeontological and neontological data. *Proceedings of the Royal Society, Biological Sciences* 274: 2413–2419.

Siliciano RF, Greene WC (2011) HIV latency. *Cold Spring Harbor Perspectives in Medicine* 1: a007096.

Simón-Soro A, Guillen-Navarro M, Mira A (2014) Metatranscriptomics reveals overall active bacterial composition in caries lesions. *Journal of Oral Microbiology* 6: 25443.

Singh H, Gallier S (2017) Nature's complex emulsion: the fat globules of milk. *Food Hydrocolloids* 68: 81–89.

Singh R, Letai A, Sarosiek K (2019) Regulation of apoptosis in health and disease: the balancing act of BCL-2 family proteins. *Nature Reviews Molecular Cell Biology* 20: 175–193.

Sinikumpu SP, Jokelainen J, Keinänen-Kiukaanniemi S, et al. (2022) Skin cancers and their risk factors in older persons: a population-based study. *BioMed Central Geriatrics* 22: 269.

Sistiaga A, Mallol C, Galván B, et al. (2014) The Neanderthal meal: a new perspective using faecal biomarkers. *Public Library of Science One* 9(6): e101045.

Skarke C, Lahens N, Rhoades S, et al. (2017) A pilot characterization of the human chronobiome. *Science Reports* 7: 17141.

Sloane C (2019) Plants with air sacs. *Sciencing.* Retrieved from: https://sciencing.com/plants-air-sacs-7471083.html. November 2, 2023.

Sluys R (2016) Invasion of the flatworms. *American Scientist* 104: 288–295.

Smith SA, Travers RJ, Morrissey JH (2015) How it all starts: initiation of the clotting cascade. *Critical Reviews in Biochemistry and Molecular Biology.* 50: 326–336.

Smith TM (2013) Teeth and human life-history evolution. *Annual Review of Anthropology* 42: 191–208.

Snyder GK, Sheafor BA (1999) Red blood cells: centerpiece in the evolution of the vertebrate circulatory system. *American Zoologist* 39: 189–198.

Song A, Severini T, Allada R (2017) How jet lag impairs Major League Baseball performance. *Proceedings of the National Academy of Sciences USA* 114: 1407–1412.

Song K, Yeom E, Lee SJ (2014) Real-time imaging of pulvinus bending in *Mimosa pudica*. *Scientific Reports* 4: 6466.

Sophia Fox AJ, Bedi A, Rodeo SA (2009) The basic science of articular cartilage: structure, composition, and function. *Sports Health* 1: 461–468.

Soto PC, Stein LL, Hurtado-Ziola N, Hedrick SM, Varki A (2010) Relative over-reactivity of human versus chimpanzee lymphocytes: Implications for the human diseases associated with immune activation. *Journal of Immunology* 184: 4185–4195.

Sotozono M, Kuriki N, Asahi Y, et al. (2021) Impacts of sleep on the characteristics of dental biofilm. *Science Reports* 11: 138.

Speller CF, Kemp BM, Wyatt SD, et al. (2010) Ancient mitochondrial DNA analysis reveals complexity of indigenous North American turkey domestication. *Proceedings of the National Academy of Sciences USA* 107: 2807–2812.

Spoor F, Garland TG, Jr, Krovitz G, et al. (2007) The primate semicircular canal system and locomotion. *Proceedings of the National Academy of Sciences USA* 104: 10808–10812.

Squier CA, Kremer MJ (2001) Biology of oral mucosa and esophagus. *Journal of the National Cancer Institute Monographs* 2001: 7–15.

Sriram K, Insel PA (2018) G Protein-coupled receptors as targets for approved drugs: how many targets and how many drugs? *Molecular Pharmacology* 93: 251–258.

Standring S (2008) Development of the limbs. In: *Gray's Anatomy* (Standring S, ed.) Churchill Livingstone, pp. 899–904

Steinbeck J (1945) *Cannery Row*. Viking Press.

Stinson LF, Boyce MC, Payne MS, et al. (2019) The not-so-sterile womb: evidence that the human fetus is exposed to bacteria prior to birth. *Frontiers in Microbiology* 10(10): 1124.

Stollar EJ, Smith DP (2020) Uncovering protein structure. *Essays in Biochemistry* 64: 649–680.

Størvold GV, Aarethun K, Bratberg GH (2013) Age for onset of walking and prewalking strategies. *Early Human Development* 89: 655–659.

Strasser A, Vaux DL (2020) Cell death in the origin and treatment of cancer. *Molecular Cell* 78: 1045–1054.

Streich AM, Todd KA (2014) *Classification and naming of plants*. University of Nebraska Lincoln. Retrieved from: https://alec.unl.edu/documents/cde/2017/natural-resources/classification-and-naming-of-plants.pdf. November 1, 2023.

Strömberg CAE, Dunn RE, Madden RH, et al. (2013) Decoupling the spread of grasslands from the evolution of grazer-type herbivores in South America. *Nature Communications* 4: 1478.

Strother PK, Foster C (2021) A fossil record of land plant origins from charophyte algae. *Science* 373: 792–796.

Stumpp M, Hu MY, Tseng Y-C, et al. (2015) Evolution of extreme stomach pH in bilateria inferred from gastric alkalization mechanisms in basal deuterostomes. *Scientific Reports* 5: 10421.

Suez J, Zmora N, Zilberman-Schapira G, et al. (2018) Post-antibiotic gut mucosal microbiome reconstitution is impaired by probiotics and improved by autologous FMT. *Cell* 174: 1406–1423.

Suman SP, Joseph P (2013) Myoglobin chemistry and meat color. *Annual Review of Food Science Technology* 4: 79–99.

Sussman MR, Phillips GN, Jr (2009) How plant cells go to sleep for a long, long time. *Science* 326: 1356–1357.

Suzanne M, Steller H (2013) Shaping organisms with apoptosis. *Cell Death Differentiation* 20, 669–675.

Swaney MH, Kalan LR (2021) Living in your skin: microbes, molecules, and mechanisms. *Infection and Immunity* 89: e00695–20.

Sweeney HL, Hammers DW (2018) Muscle contraction. *Cold Spring Harbor Perspectives in Biology* 10: a023200.

Sweis R, Fox M (2020) The global burden of gastro-oesophageal reflux disease: more than just heartburn and regurgitation. *Lancet Gastroenterology and Hepatology* 5: 519–521.

Talbot G (2017) *Oils and fats.* Institute of Food Science Technology. Retrieved from: https://www.ifst.org/resources/information-statements/oils-and-fats. November 2, 2023.

Taylor RG, Scanlon B, Döll P, et al. (2013) Groundwater and climate change: recent advances and a look forward. *Nature Climate Change* 3: 322–329.

Teaford MF, Ungar PS (2000) Diet and the evolution of the earliest human ancestors. *Proceedings of the National Academy of Sciences USA* 97: 13506–13511.

Tettamanti G, Casartelli M (2019) Cell death during complete metamorphosis. *Philosophical Transactions of the Royal Society, Part B* 374: 20190065.

Thaler EA, Kwang JS, Quirk BJ, et al. (2022) Rates of historical anthropogenic soil erosion in the Midwestern United States. *Earth's Future* 10: e2021EF002396.

Thaler EA, Larsen IJ, Yu Q (2021) The extent of soil loss across the US Corn Belt. *Proceedings of the National Academy of Sciences USA* 118(8): e1922375118.

Thiebes K (2021) What is blood made of? Review blood components and functions. *Simplified Science.* Retrieved from: https://www.simplifiedsciencepublishing.com/resources/what-is-blood-made-of-and-review-of-human-blood-components-and-functions.   November 1, 2023.

Thoreau HD (1849) *A Week on the Concord and Merrimack Rivers.* Princeton University Press.

Thorstenson CA, Pazda AD, Krumhuber EG (2021) The influence of facial blushing and paling on emotion perception and memory. *Motivation and Emotion* 45: 818–830.

Tian F, Stevens NM, Buckler ES (2009) Tracking footprints of maize domestication and evidence for a massive selective sweep on chromosome 10. *Proceedings of the National Academy of Sciences USA* 106(Suppl 1): 9979–9986.

Tidwell JH, Allan GL (2001) Fish as food: aquaculture's contribution: ecological and economic impacts and contributions of fish farming and capture fisheries. *EMBO Reports* 2: 958–963.

Tilly JL (2001) Commuting the death sentence: how oocytes strive to survive. *Nature Reviews Molecular Cell Biology* 2: 838–848.

Tiwari M, Prasad S, Tripathi A, et al. (2015) Apoptosis in mammalian oocytes: a review. *Apoptosis* 20: 1019–1025.

Tolpadi AA, Lee JJ, Pedoia V, et al. (2020) Deep learning predicts total knee replacement from magnetic resonance images. *Scientific Reports* 10: 6371.

Totsche KU, Amelung W, Gerzabek MH, et al. (2018) Microaggregates in soils. *Journal of Plant Nutrition and Soil Science* 181: 104–136.

Tower J (2015) Programmed cell death in aging. *Ageing Research Reviews* 23(Pt A): 90–100.

Travaglini KJ, Nabhan AN, Penland L, et al. (2020) A molecular cell atlas of the human lung from single-cell RNA sequencing. *Nature* 587: 619–625.

Traylen CM, Patel HR, Fondaw W, et al. (2011) Virus reactivation: a panoramic view in human infections. *Future Virology* 6: 451–463.

Truman JW (2019) The evolution of insect metamorphosis. *Current Biology* 29: R1252–R1268.

Tsujimoto Y, Shimizu S (2005) Another way to die: autophagic programmed cell death. *Cell Death and Differentiation* 12: 1528–1534.

Tsunetsugu Y, Park BJ, Miyazaki Y (2010) Trends in research related to "Shinrin-yoku" (taking in the forest atmosphere or forest bathing) in Japan. *Environmental Health and Preventive Medicine* 15: 27. https://doi.org/10.1007/s12199-009-0091-z.

Turnbaugh P, Ley R, Hamady M, et al. (2007) The Human Microbiome Project. *Nature* 449: 804–810.

Turner AH, Makovicky PJ, Norell MA (2007) Feather quill knobs in the dinosaur *Velociraptor. Science* 317: 1721.

Ueda M, Nakamura Y (2007) Chemical basis of plant leaf movement. *Plant and Cell Physiology* 48: 900–907.

Ursell LK, Metcalf JL, Parfrey LW, et al. (2012) Defining the human microbiome. *Nutrition Review* 70(Suppl 1): S38–S44.

van den Hoogen J, Geisen S, Routh D, et al. (2019) Soil nematode abundance and functional group composition at a global scale. *Nature* 572: 194–198.

van der Meijden PEJ, Heemskerk JWM (2019) Platelet biology and functions: new concepts and clinical perspectives. *Nature Reviews Cardiology* 16: 166–179.

van Doorn WG, van Meeteran U (2003) Flower opening and closure: a review. *Journal of Experimental Botany* 54: 1801–1812.

van Haeringen NJ (2001) The (neuro) anatomy of the lacrimal system and the biological aspects of crying. In: *Adult Crying: Psychological and Psychobiological Aspects* (Vingerhoets A, Cornelius R, eds). Brunner-Routledge, pp. 19–36.

van Kijk JG (2003) Fainting in animals. *Clinical Autonomic Research* 13: 247–255.

van West P, Morris BM, Reid B, et al. (2002) Oomycete plant pathogens use electric fields to target roots. *Molecular Plant-Microbe Interactions* 15: 790–798.

Varki A, Altheide TK (2005) Comparing the human and chimpanzee genomes: searching for needles in a haystack. *Genome Research* 15: 1746–1758.

Varki A, Geschwind DH, Eichler EE (2008) Explaining human uniqueness: genome interactions with environment, behaviour and culture. *Nature Reviews Genetics* 9: 749–763.

Veiga P, Suez J, Derrien M, et al. (2020) Moving from probiotics to precision probiotics. *Nature Microbiology* 5: 878–880.

Venkat A, Muneer S (2022) Role of circadian rhythms in major plant metabolic and signaling pathways. *Frontiers in Plant Science* 13: 836244. https://doi.org/10.3389/fpls.2022.836244.

Verma AK, Prakash S (2020) Status of animal phyla in different kingdom systems of biological classification. *International Journal of Biological Innovations* 2: 149–154.

Vickers NJ (2017) Animal communication: when I'm calling you, will you answer too? *Current Biology* 27: R713–R715.

Vilanova XM (2014) Shellfish. In: *Encyclopedia of Meat Sciences*, 2nd edition (Dikeman M and Devine C, eds). Academic Press, pp. 380–387.

Vilgis TA (2015) Soft matter food physics—the physics of food and cooking. *Reports on Progress in Physics* 78(12): 124602.

Vitale Brovarone A, Sverjensky DA, Piccoli F, et al. (2020) Subduction hides high-pressure sources of energy that may feed the deep subsurface biosphere. *Nature Communications* 11: 3880.

Vogt JT, Smith WB (2016) *Forest Inventory and Analysis: fiscal Year 2016 Business Report*. U.S. Department of Agriculture. Retrieved from: https://www.fs.usda.gov/sites/default/files/fs_media/fs_document/publication-15817-usda-forest-service-fia-annual-report-508.pdf. November 1, 2023. pp. 71–72.

Vosshall LB, Price JL, Sehgal A, et al. (1994) Block in nuclear localization of *period* protein by a second clock mutation, *timeless*. *Science* 263: 1606–1609.

Vyazovskiy VV, Harris KD (2013) Sleep and the single neuron: the role of global slow oscillations in individual cell rest. *Nature Reviews Neuroscience* 14: 443–451.

Wagg C, Bender SF, Widmer F, et al. (2014) Soil biodiversity and soil community composition determine ecosystem multifunctionality. *Proceedings of the National Academy of Sciences USA* 111: 5266–5270.

Wagner SC (2011) Biological nitrogen fixation. *Nature Education Knowledge* 3(10): 15.

Waksman S (1952) *Soil Microbiology*. Wiley.

Walker MP (2008) Cognitive consequences of sleep and sleep loss. *Sleep Medicine* 9: S29–S34.

Walker MP (2021) Sleep essentialism. *Brain* 144: 697–699.

Wallace JL (2008) Prostaglandins, NSAIDs, and gastric mucosal protection: why doesn't the stomach digest itself? *Physiology Review* 88: 1547–1565.

Walther TC, Chung J, Farese RV, Jr (2017) Lipid droplet biogenesis. *Annual Review of Cell and Developmental Biology* 33: 491–510.

Ward P (2009) What will become of *Homo sapiens*? *Scientific American* 300: 68–73.

Ward ZJ, Bleich SN, Cradock AL, et al. (2019) Projected U.S. state-level prevalence of adult obesity and severe obesity. *New England Journal of Medicine* 381: 2440–2450.

Wassenaar TM, Panigrahi P (2014) Is a foetus developing in a sterile environment? *Letters in Applied Microbiology* 59: 572–579.

Waters CM, Bassler BL (2005) Quorum sensing: cell-to-cell communication in bacteria. *Annual Review of Cell and Developmental Biology* 21: 319–346.

Watson H (2015) Biological membranes. *Essays in Biochemistry* 59: 43–69.

Watve M, Parab S, Jogdand P, et al. (2006) Aging may be a conditional strategic choice and not an inevitable outcome for bacteria. *Proceedings of the National Academy of Sciences USA* 103: 14831–14835

Wayman E (2011) Six talking apes. *Smithsonian Magazine*. Retrieved from: https://www.smithsonianmag.com/science-nature/six-talking-apes-48085302/. November 2, 2023.

Weems AD, Welf ES, Driscoll MK, et al. (2023) Blebs promote cell survival by assembling oncogenic signalling hubs. *Nature* 615: 517–525.

Weinberg JM (2007) Herpes zoster: epidemiology, natural history, and common complications. *Journal of the American Academy of Dermatology* 57: S130–S135.

Weisshaar E (2016) Epidemiology of itch. *Current Problems in Dermatology* 50: 5–10.

Weitbrecht K, Müller K, Leubner-Metzger G (2011) First off the mark: early seed germination. *Journal of Experimental Botany* 62: 3289–3309.

Wenner Moyer M (2013) Whole-grain foods not always healthful. *Scientific American*. Retrieved from: https://www.scientificamerican.com/article/whole-grain-foods-not-always-healthful/. November 1, 2023.

Weschler CJ, Langer S, Fischer A, et al. (2011) Squalene and cholesterol in dust from Danish homes and daycare centers. *Environmental Science and Technology* 45: 3872–3879.

Weyrich LS, Duchene S, Soubrier J, et al. (2017) Neanderthal behaviour, diet, and disease inferred from ancient DNA in dental calculus. *Nature* 544: 357–361.

Whitley RJ, Kimberlin DW, Roizman B (1998) Herpes simplex viruses. *Clinical Infectious Diseases* 26: 541–553.

Whitley RJ, Roizman B (2001) Herpes simplex virus infections. *Lancet* 357: 1513–1518.Willett WC, Ludwig DS (2020) Milk and health. *New England Journal of Medicine* 382: 644–654.

Whitman WB, Coleman DC, Wieve WJ (1998) Prokaryotes: the unseen majority. *Proceedings of the National Academy of Sciences USA* 95: 6578–6583.

Wilson EO (1975) *Sociobiology: The New Synthesis*. Harvard University Press.

Wilson EO (1997) Biophilia and the conservation ethic. In: *The Biophilia Hypothesis* (Kellert SR and Wilson EO, eds). Island Press, p. 31.

Wilson M, Wilson TP (2005) An oscillator model of the timing of turn-taking. *Psychonomic Bulletin and Review* 12: 957–968.

Wilsterman K, Ballinger MA, Williams CM (2021) A unifying, ecophysiological framework for animal dormancy. *Functional Ecology* 35: 11–31.

Winans JB, Wucher BR, Nadell CD (2022) Multispecies biofilm architecture determines bacterial exposure to phages. *Public Library of Science Biology* 20(12): e3001913.

Windt JM (2021) How deep is the rift between conscious states in sleep and wakefulness? Spontaneous experience over the sleep–wake cycle. *Philosophical Transactions of the Royal Society* 376: 20190696.

Winkler LL, Christensen U, Glümer C, et al. (2016) Substituting sugar confectionery with fruit and healthy snacks at checkout—a win-win strategy for consumers and food stores? A study on consumer attitudes and sales effects of a healthy supermarket intervention. *BioMed Central Public Health* 16: 1184.

Winslow RM (2007) Blood substitutes: Basic principles and practical aspects. In: *Blood Banking and Transfusion Medicine: Basic Principles and Practice* 2nd ed. (Hillyer CD, Silberstein LE, Ness PM, Anderson KC, Roback JD, eds). Churchill Livingston, Elsevier.

Winson J (1990) The meaning of dreams. *Scientific American* 263: 42–48.

Winter WC, Hammond WR, Green NH, et al. (2009) Measuring circadian advantage in Major League Baseball: a 10-year retrospective study. *International Journal of Sports Physiology and Performance* 4: 394–401.

Wittig K, Kasper J, Seipp S, et al. (2011) Evidence for an instructive role of apoptosis during the metamorphosis of *Hydractinia echinata* (Hydrozoa). *Zoology* 114: 11–22.

Wong RS (2011) Apoptosis in cancer: from pathogenesis to treatment. *Journal of Experimental and Clinical Cancer Research* 30: 87.

Wood JD, Enser M, Fisher AV, et al. (2008) Fat deposition, fatty acid composition and meat quality: a review. *Meat Science* 78: 343–358.

Worchel FF, Allen MA (1997) Mothers' ability to discriminate cry types in low-birthweight premature and full-term infants. *Children's Health Care* 26: 183–195.

Worley AC, Raper KB, Hohl M (1979) *Fonticula alba*: a new cellular slime mold (Acrasiomycetes). *Mycologia* 71: 746–760.

Wrangham R (2017) Control of fire in the Paleolithic: evaluating the cooking hypothesis. *Current Anthropology* 58: S303–S313.

Wrangham RW, Jones JH, Laden G, Pet al. (1999) The raw and the stolen: cooking and the ecology of human origins. *Current Anthropology* 40: 567–594.

Xu L, Li W, Voleti V, et al. (2020) Widespread receptor-driven modulation in peripheral olfactory coding. *Science* 368: eaaz5390.

Yamaguchi Y, Miura M (2015) Programmed cell death in neurodevelopment, *Developmental Cell* 32: 478–490.

Yang J, McCormick MA, Zheng J, et al. (2015) Systematic analysis of asymmetric partitioning of yeast proteome between mother and daughter cells reveals "aging factors" and mechanism of lifespan asymmetry. *Proceedings of the National Academy of Sciences USA* 112: 11977–11982.

Yokouchi M, Atsugi T, van Logtestijn M, et al. (2016) Epidermal cell turnover across tight junctions based on Kelvin's tetrakaidecahedron cell shape. *eLife* 5: e19593.

Young LS, Rickinson AB (2004) Epstein-Barr virus: 40 years on. *Nature Reviews Cancer* 4: 757–768.

Young RW (2003) Evolution of the human hand: the role of throwing and clubbing. *Journal of Anatomy* 202: 165–174.

Yousef H, Alhajj M, Sharma S (2023) *Anatomy, skin (integument), epidermis. [Updated November 14, 2022]. In: StatPearls [Internet].* StatPearls Publishing. Retrieved from: https://www.ncbi.nlm.nih.gov/books/NBK470464/. October 23, 2023.

Yuan C-X, Ji Q, Meng Q-J, Tabrum AR, Zhe-Xi Luo Z-X (2013) Earliest evolution of multituberculate mammals revealed by a new Jurassic fossil. *Science* 341: 779–783.

Yuan J, Yankner BA (2000) Apoptosis in the nervous system. *Nature* 407: 802–809.

Zelik KE, Adamczyk PG (2016) A unified perspective on ankle push-off in human walking. *Journal of Experimental Biology* 219: 3676–3683.

Zhang YJ, Li S, Gan RY, et al. (2015) Impacts of gut bacteria on human health and diseases. *International Journal of Molecular Science* 16: 7493–7519.

Zhang Z (2016) Mechanics of human voice production and control. *Journal of Acoustical Society of America* 140(4): 2614.

Zhao X, Zhang J, Zhu KY (2019) Chito-protein matrices in arthropod exoskeletons and peritrophic matrices. In: *Extracellular Sugar-Based Biopolymers Matrices* (Cohen E and Merzendorfer H, eds). Springer, pp. 3–56.

Zhou Z, Xu MJ, Gao B (2016) Hepatocytes: a key cell type for innate immunity. *Cell and Molecular Immunology* 13: 301–315.

Zimecki M (2006) The lunar cycle: effects on human and animal behavior and physiology. *Postępy Higieny i Medycyny Doświadczalnej* (Online) 60: 1–7. Retrieved from: https://pub med.ncbi.nlm.nih.gov/16407788/#:~:text=Animal%20studies%20revealed%20that%20 the,disappear%20during%20full%2Dmoon%20days. November 2, 2023.

Zink KD, Lieberman DE (2016) Impact of meat and Lower Palaeolithic food processing techniques on chewing in humans. *Nature* 531: 500–503.

Zlinszky A, Molnár B, Barfod AS (2017) Not all trees sleep the same—high temporal resolution terrestrial laser scanning shows differences in nocturnal plant movement. *Frontiers in Plant Science* 8: e1814.

Zrzavý J, Štys P (1997) The basic body plan of arthropods: insights from evolutionary mor-phology and developmental biology. *Journal of Evolutionary Biology* 10: 353–367.

Index

For the benefit of digital users, indexed terms that span two pages (e.g., 52–53) may, on occasion, appear on only one of those pages.

Tables and figures are indicated by an italic *t* and *f* following the page number.